U. Schwertmann, R. M. Cornell

Iron Oxides in the Laboratory

Second, Completely Revised and Extended Edition

WILEY-VCH

U. Schwertmann
R. M. Cornell

Iron Oxides in the Laboratory

Preparation and Characterization

Second, Completely Revised and Extended Edition

WILEY-VCH

Weinheim · New York · Chichester
Brisbane · Singapore · Toronto

Professor Dr. Dr. h. c. Udo Schwertmann
Lehrstuhl für Bodenkunde
Technische Universität München
D-85350 Freising-Weihenstephan
Federal Republic of Germany

Dr. Rochelle M. Cornell
Universität Bern
Departement für Chemie und Biochemie
CH-3000 Bern 9
Switzerland

First Edition 1991
Second, Completely Revised and Extended Edition 2000

Library of Congress Card No. applied for

British Library Cataloguing-in-Publication Data

Deutsche Nationalbibliothek Cataloguing-in-Publication Data
A catalogue record for this publication is available from the Deutsche Nationalbibliothek
ISBN 978-3-527-29669-9

Composition: ProSatz Unger, Weinheim

Printed and bound by CPI Group (UK) Ltd, Croydon, CR0 4YY

C9783527296699_071125

Preface to the Second Edition

Since this book first appeared, there has been a surge of interest in synthetic fine to ultrafine iron oxides in a wide range of scientific and technical disciplines, especially in mineralogy, the geosciences and environmental science and in various branches of technology. In this work, the recipes of this book have been frequently used. However, a great deal of new information has been produced with respect to iron oxide synthesis, particularly regarding ultrafine iron oxides and those with different morphologies. We have taken these developments into account in the second edition.

As before, our main aim was to provide reliable, well tested recipes for the synthesis of pure iron oxides. In addition, all chapters have been revised and new data added. In the initial, general chapters we have expanded the section on monodispersed particles, currently of great interest to industry and offered some illustration of such products. The methods of characterization have been focused on their relevance to iron oxides and for general aspects of these methods, the reader is referred to standard treatises. The chapter on color, which is of paramount importance for iron oxides, has been enlarged to incorporate recent advances in this field.

In the recipe chapters, we have retained the well tried syntheses from our own laboratories and added some more, recent ones. In addition, we refer, although not in great detail, to a number of procedures developed in other laboratories, which we have not reproduced ourselves. The aim here was to extend the scope of the book. Where we describe the product and comment on its formation, much new information has been added to provide a better understanding of possible variations of the product and the influence of the formation conditions on these variations.

We have added two new chapters. One describes the preparation of iron-oxide-coated SiO_2 which was developed by A. Scheidegger in 1993. This material is increasingly used for studies of a range of reactions with iron oxides in percolation experiments in columns. The other chapter is a lecture/demonstration of the formation of the most common iron oxides.

We hope that our new edition will continue to be of use to all those who want to produce well-defined iron oxides with a range of properties for scientific investigations.

As in the first edition, we acknowledge the active assistance of a large number of colleagues from around the world. Dr. A. Scheinost, University of Delaware, supplied the chapter on color and Dr. A. Scheidegger, P. Scherrer Institut, Zürich, revised the chapter on coating SiO_2. Critical reading of different chapters was provided by Drs. E. Murad, Bayrisches Geologisches Landesamt, Bamberg; Drs. H. Stanjek and J. Friedl, Institut für Bodenkunde, TU München. Mr. H. Fechter from the same institute worked out the experimental part of the demonstration lecture. Dr. Murad also produced the Mössbauer spectra and Dr. H.-Ch. Bartscherer, Physikalisches Institut, TU München, produced most of the electron micrographs. Photographic material was kindly supplied by Prof. T. Sugimoto, Tohuko, University; Dr. D. Janney, University of Arizona; Dr. P. Weidler, ETH Zürich; Dr. S. Glasauer, University of Guelph and Dr. R. Giovanoli, Universität Bern. We are also grateful to Drs. G. Buxbaum and P. Woditsch, Bayer AG, Leverkusen and Krefeld, for providing Plate I and the cover picture; Dr. L. Carlson, University of Helsinki, for Plate II and Dr. J. M. Bigham, Ohio State University, for Plate VIII. One of us (R.M.C.) wants to thank her mother for being an unfailing source of inspiration and encouragement during the preparation of this book. Our warmest thanks go to all of these people.

Finally, we should like to thank the staff of Wiley-VCH for their patience and cooperation in the production of this book.

April 2000

U. Schwertmann
R. M. Cornell

Contents

Introduction

Iron oxides and oxide hydroxides, for the sake of simplicity all of these are referred to as "iron oxides", play an important role in a variety of disciplines, including pure, environmental and industrial chemistry, corrosion science, mineralogy, geology, soil science, planetology, biology and medicine. Workers in all these fields frequently need to prepare synthetic iron oxides. These iron oxides may be required for investigations of their own particular properties or used as starting materials for other processes. They also serve as model systems for investigation of dissolution mechanisms, adsorption of ions and molecules and reduction and catalysis reactions.

There are four main fields where Fe oxides are of interest. These fields and their predominant Fe oxides are:

1) *Soil science, mineralogy, geology:* goethite, hematite, lepidocrocite, ferrihydrite, maghemite and magnetite, akaganeite, schwertmannite, feroxyhite
2) *Corrosion science:* magnetite, maghemite, goethite, lepidocrocite, wüstite and green rusts
3) *Industry:* goethite, hematite, magnetite (inorganic pigments; Plate I), maghemite (magnetic tapes), goethite, hematite, ferrihydrite (clays and ceramic materials), ferrihydrite, hematite (catalysts)
4) *Biology:* Ferrihydrite (ferritin, iron overload), lepidocrocite, goethite, magnetite (biominerals), goethite, ferrihydrite, lepidocrocite (dissimilatory respiration)

For all these purposes reliable methods of producing pure iron oxides are required. This book aims at fulfilling this need. It sets out reliable, convenient preparative methods for synthesizing the major iron oxides. The emphasis is on techniques and equipment that are readily accessible to the average scientist. For this reason and also to meet the interests of those geo- and bioscientists wishing to work with synthetic Fe oxides whose properties are close to those of natural ones, the preparations usually involve precipitation or transformation in aqueous solution at temperatures lower than 100 °C. A heating oven or hot plate, a pH meter, an accurate balance and in some cases, a source of N_2, air or O_2 together with an automatic titrimeter are the only pieces of equipment usually required.

The book opens with a short introduction into the nature of iron oxides. This is followed by a discussion of general preparative techniques (chapter 2). In chapter 3, techniques for characterization of the products – color measurement, electron microscopy, X-ray diffraction, infra red absorption spectroscopy, surface area measurement, thermoanalysis and Mössbauer spectroscopy – are briefly described with particular emphasis on their application to Fe oxides.

The main chapters (4–14) are concerned with the preparative methods. For the majority of oxides particularly hematite and goethite more than one preparative method is described. Properties such as crystal morphology and surface area frequently depend on preparative conditions and a selection of methods is presented to enable a range of oxides with specific characteristics to be produced. The production of so-called monodisperse Fe oxides, i. e. products with a rather narrow particle size distribution, is also included.

In natural Fe oxides the octahedrally coordinated Fe^{III} ion is often partially replaced by other metal cations especially by Al. These substituted Fe oxides have attracted wide interest in recent research, so methods for preparing partly substituted Fe oxides are also included here.

The detailed description of each preparative procedure is followed by a description of the properties of the product and by comments concerning possible variations of the method and their effects on the properties of the product. For thorough characterization numerous illustrations including color plates, X-ray diffractograms, absorption spectra (IR, Mössbauer) and electron micrographs are included. This characterization is necessary in view of the wide range of crystal morphologies and crystal sizes displayed by most iron oxides. It should enable the users of this book to obtain a particular product with the desired characteristics and also provides a check on the success of the users' own efforts.

Most of preparative methods provided have been tested in our laboratories. In order to widen the scope of these chapters, several other methods of preparing the various iron oxides, not tested by us, are also briefly described. For details the user is referred to the appropiate reference. The method of preparing iron oxide-coated SiO_2, developed by Scheidegger et al. (1994), is also included. The final chapter consists of a lecture in which the synthesis of several Fe oxides is demonstrated at the bench.

The book concludes with a bibliography. This is not intended to be exhaustive. It refers mainly to the methods described here and provides further details for any reader who wishes to investigate the subject in more depth. Where possible the reference is to the original work, not to a review.

1 The Iron Oxides

This chapter is intended to offer a very brief introduction to the most important general features of iron oxides. It deals with the forms, chemistry, crystal structure, important properties and formation of these compounds and their enviromental and industrial significance. For more information the reader is referred to the monograph "The Iron Oxides" by the same authors published in 1996 by Wiley-VCH.

There are fifteen iron oxides, oxide hydroxides and hydroxides known to date. The more important ones are listed in Table 1. In addition to these, $Fe(OH)_2$, FeO (wüstite), a β-Fe_2O_3, a ε-Fe_2O_3, a high pressure FeOOH, a ferrimagnetic δ-FeOOH and a crystalline $Fe(OH)_3$ (bernalite) exist. Except for ferrihydrite and feroxyhyte, these compounds can be obtained in well crystallized form*. There is also a group of Fe^{III}-oxy-hydroxy salts which are closely related to the pure oxides. To this group belong an Fe^{III} oxyhydroxy sulfate (schwertmannite), and an oxyhydroxy nitrate. The chloride form is in fact the mineral akaganeite, although usually written as β-FeOOH. Other halogenides can also be incorporated into akaganeite. Fin-

Table 1-1. The Major Iron Oxides and Oxide Hydroxides

Oxyhydroxides		Oxides	
Formula	Mineral	Formula	Mineral
α-FeOOH	Goethite	$Fe_5HO_8 \cdot 4H_2O$	Ferrihydrite
β-FeOOH	Akaganeite	α-Fe_2O_3	Hematite
γ-FeOOH	Lepidocrocite	γ-Fe_2O_3	Maghemite
δ'-FeOOH	Feroxyhyte	Fe_3O_4	Magnetite

* To date, bernalite has not been synthesized in the laboratory

ally, there is a group of $Fe^{II,III}$ hydroxy salts, the green rusts, so-called because of their green-blue color and their occurrence as corrosion products.

All iron oxides and hydroxides consist of Fe, O and/or OH. They differ in composition, in the valence of Fe and, above all, in crystal structure.

1.1 The Major Iron Oxides

The basic structural unit of all Fe^{III} oxides is an octahedron, in which each Fe atom is surrounded either by six O or by both O and OH ions. The O and OH ions form layers which are either approximately hexagonally close-packed (hcp), as in goethite and hematite, or approximately cubic close-packed (ccp), as in lepidocrocite and maghemite. In both hcp and ccp stuctures, tetrahedral interstices also exist between three O or OH in one plane and the anion in the plane above. The two hcp forms, goethite and hematite, are termed α-phases, whereas the corresponding ccp forms, lepidocrocite and maghemite, are termed γ-phases.

The Fe^{3+} in the octahedral position may be partly replaced by other trivalent metal cations of similar size, for example, Al^{3+}, Mn^{3+}, Cr^{3+} and V^{3+}, without modifying the structure (*isomorphous substitution*). In this way solid solutions between pure end members (such as FeOOH and AlOOH) are formed. Other cations, e.g. Ni^{2+}, Co^{3+}, Zn^{2+}, Cd^{2+}, Pb^{4+}, and Cu^{2+} can also be incorporated into the Fe oxide structure, paticularly into goethite (Gerth, 1990; Giovanoli and Cornell, 1992; Lim-Nunez and Gilkes, 1987; Sidhu et al. 1978). A list of those elements which have been incorporated into various iron oxides is given in Table 1-2 (for a

Table 1-2. Ionic Radii of Fe and Metal Substituents

Ion	Fe^{2+}	Fe^{3+}	Al^{3+}	Cr^{3+}	Mn^{3+b}	V^{3+}	Co^{3+}	Cu^{2+b}	Zn^{2+}	Pb^{4+}	Cd^{2+}
Radius[a]/nm	0.077	0.064	0.053	0.061	0.065	0.064	0.053	0.073	0.075	0.078	0.095

[a] Ionic radius depends on whether the ion is in the high spin or low spin state. In the Fe oxide structure (and where a choice exists), the ions listed adopt a high spin state.

[b] These ions display the Jahn-Teller effect which leads to distortion of the coordination sphere of the ion.

complete table and further references see Cornell and Schwertmann, 1996).

The various Fe oxides differ mainly in the arrangement of the $Fe(O,OH)_6$ octahedra.

Goethite, lepidocrocite and akaganeite, i. e. the FeOOH forms, all consist of double bands of edge-sharing $FeO_3(OH)_3$ octahedra (Figure 1-1). In *goethite*, as in diaspore (α-AlOOH) the double bands are linked by corner-sharing in such a way as to form 2×1 octahedra "tunnels" crossed by hydrogen bridges. Whereas these "tunnels" are only large enough to permit the passage of protons, the structure of the second FeOOH form, *akaganeite* (β-FeOOH), contains squared molecular channels bounded by four double rows of octahedra and large enough to accomodate anions. In akaganeite these channels are stabilized by being filled with variable amounts of chloride anions (Naono, 1993, suggested that H_2O molecules fill the channel whereas the Cl belongs in the coordination shell of the Fe).

Two poorly ordered compounds both having the akaganeite structure but with sulfate or nitrate instead of chloride in the tunnel have been found recently. The sulfate form occurs frequently in nature as an oxidation product of pyrite and has, therefore, been recognized as a mineral with the name schwertmannite (Bigham et al. 1994). Chemically it can be considered a Fe^{III}oxyhydroxy sulfate with the ideal formula $Fe_8O_8(OH)_6SO_4$. The corresponding nitrate form can be synthesized by forced hydrolysis of an acidic $Fe^{III}(NO_3)_3$ solution at 80°C and is a precursor of ferrihydrite (Schwertmann et al. 1996).

In *lepidocrocite* as in boehmite (γ-AlOOH), the double bands of octahedra share edges to form zig-zag layers which are connected to each other by hydrogen bonds (OH…O).

Hematite consists of layers of FeO_6 octahedra which are connected by edge- and face-sharing (as in corundum, α-Al_2O_3) and stacked perpendicular to the *c* direction. Two thirds of the octahedral interstices are filled with Fe^{III}. The face-sharing is accomplished by a slight distortion of the octahedra which causes a regular displacement of the Fe ions. The distortion and the absence of H-bonds yield a compact structure which is responsible for the high density of 5.26 g/cm^3. Hematites synthesized at low temperatures in an aqueous medium usually contain some OH in the structure (Wolska, 1981).

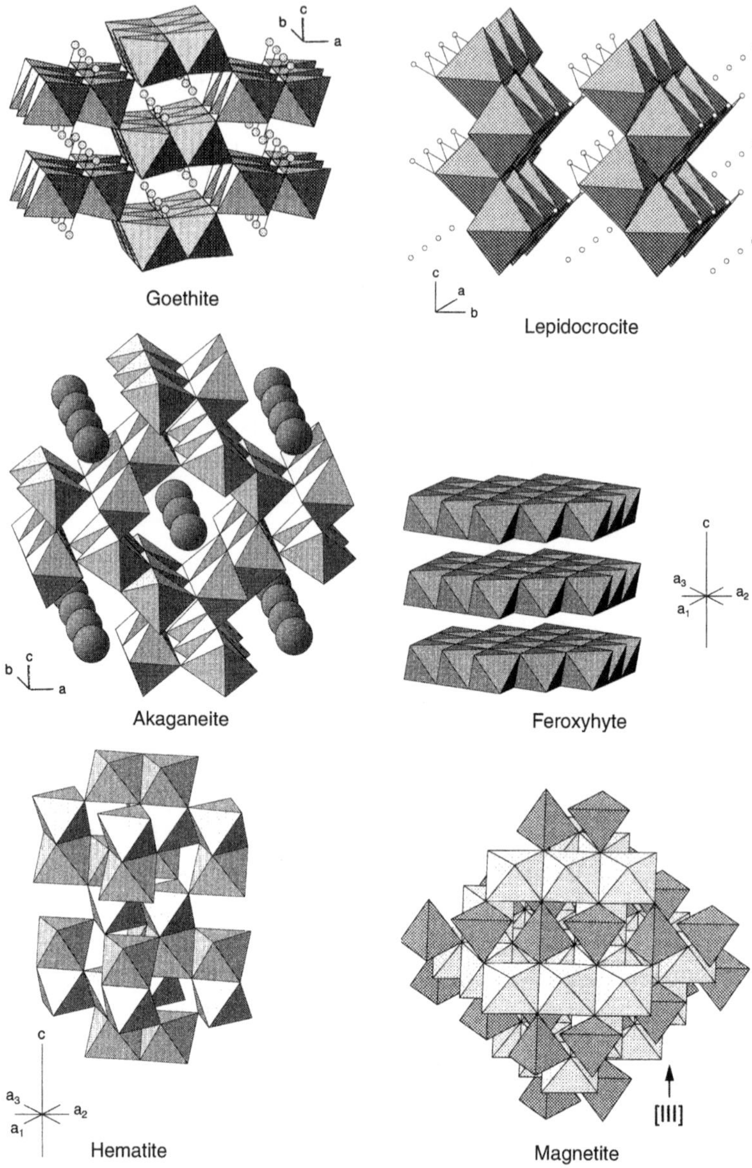

Fig. 1-1. Structural models of goethite, akaganeite, lepidocrocite, feroxyhyte, hematite and magnetite. Small, open circles indicate H atoms (from H. Stanjek, in Cornell and Schwertmann, 1996, with permission).

In the cubic structure of both *magnetite* and *maghemite* 1/3 of the interstices are tetrahedrally coordinated with oxygen and 2/3 are octahedrally coordinated. In magnetite all of these positions are filled with Fe. Magnetite is an inverse spinel: the tetrahedral positions are completely occupied by Fe^{III}, the octahedral ones by equal amounts of Fe^{III} and Fe^{II}. In maghemite only 5/6 of the total available positions are filled and only by Fe^{3+}, the remainder are vacant (□) (i.e. $Fe_{2.67}\square_{0.33}O_4$). Maghemite can have different symmetries depending on the degree of ordering of the vacancies (Bernal et al., 1959). Completely ordered maghemite has a primitive cubic or a tetragonal cell and the X-ray pattern shows "superstructure lines" (Fig. 13-3); otherwise it is cubic. Maghemite can, thus, be considered a fully oxidized magnetite. Oxidation is achieved through the ejection of 11% of the Fe from the structure thereby creating the vacancies. The color changes from black to red-brown. A complete series of transitions exists between magnetite and maghemite.

Although the structures of *ferrihydrite* and *feroxyhyte* (δ'-FeOOH) are not yet fully understood, both compounds are considered to have a predominately hematite-like structure consisting of hexagonally close packed oxygen planes with Fe ions in the octahedral interstices. The periodicity of the octahedral sheets along the z-direction is 2 for feroxyhite ($c = 0.46$ nm), 4 for ferrihydrite ($c = 0.94$ nm) and 6 for hematite ($c = 1.3752$ nm). The low degree of crystallinity of ferrihydrite and feroxyhyte is linked both to the presence of vacant Fe sites in the structure and to replacement of some oxygen by H_2O and/or OH. For ferrihydrite, a range of compounds with different degrees of structural order exist; these compounds are generally named according to the number of broad X-ray peaks which they exhibit: e.g. 2-line and 6-line ferrihydrite. The 2-line ferrihydrite is also called protoferrihydrite (Chukhrov et al., 1973). The formula of ferrihydrite has not been fully established. $Fe_5HO_8 \cdot 4\,H_2O$ has been suggested, but the structural water may in fact be less than indicated in this formula.

Green Rusts are not oxides or hydroxides in a strict sense but contain anions as an essential structural component. They consist of hexagonally close-packed layers of O and OH of the $Fe(OH)_2$-structure type with Fe^{II} and Fe^{III} in the interstices (Feitknecht and Keller, 1950). Fe^{III} gives the layer a positive charge which is balanced by intercalation of anions, such as chloride, sulfate, carbonate etc., between the layers. The sulfate and chloride forms have a maximum Fe^{II}/Fe^{III} ratio of 4 if completely unoxidized, whereas the carbonate form has a ratio of 2 or 3. The general structure of the green rusts can be written as $[Fe_a^{III}Fe_b^{II}(OH)_c]$ $[A_{(3a+2b-c)/z}] \cdot H_2O$, where A represents the interlayer anion of valence z (Lewis. 1997).

Inter-conversions between the compounds listed in Table 1-1 are possible and often occur readily. The oxide hydroxides dehydrate to their anhydrous structural counterparts and ultimately to hematite on heating. At lower temperatures and in solution, inter-conversions usually require dissolution followed by reprecipitation of the new phase. Oxidation or reduction reactions are also possible. An overview of the more frequent inter-conversions is presented in Fig. 1-2.

Physical and chemical properties of the major iron oxides are summarized in Table 1-3.

1.2 Less Common or Rare Iron Oxides

The minor iron oxides are $Fe(OH)_2$, FeO (wüstite) β-Fe_2O_3, ε-Fe_2O_3 and the high pressure form of FeOOH. The latter three compounds have not been found in nature.

$Fe(OH)_2$ is prepared from Fe^{II} solutions by precipitation with alkali. When freshly precipitated under an inert atmosphere (in a Schlenck apparatus for example) $Fe(OH)_2$ is white (Bernal et al., 1959). It is, however, readily oxidized by air or even water upon which it darkens. $Fe(OH)_2$ has the CdI_2 type structure with hcp anions and half of the octahedral interstices being filled with Fe^{2+} ions. The crystals form hexagonal platelets. In solution $Fe(OH)_2$ transforms by a combination of oxidation/de-hydration/hydrolysis reactions to other iron oxides and hydroxides. The end product depends both upon the order in which these processes occur and upon their rates.

A *high pressure synthetic form of FeOOH* can be prepared by hydrothermal conversion of α-Fe_2O_3 in NaOH at 500 °C at a pressure of 80–90 kb for one hour (Chenavas et al., 1973). The crystal structure is that of InOOH. It consists of single chains of octahedra running parallel to the c axis and linked by hydrogen bonds.

Plate I. Industrial production of goethite pigment. (Courtesy: Bayer AG, Krefeld, Germany)

Plate II. Iron oxide formation in the environment: Ferrihydrite formed by oxidation of Fe^{2+} in ferriferous spring in Iceland. (Courtesy: L.Carlson, University of Helsinki).

Plate III. Schwertmannite formation in an acid mine water stream in Ohio (Courtesy J.M. Bigham, Ohio State University).

Plate IV. The Munsell color tree.

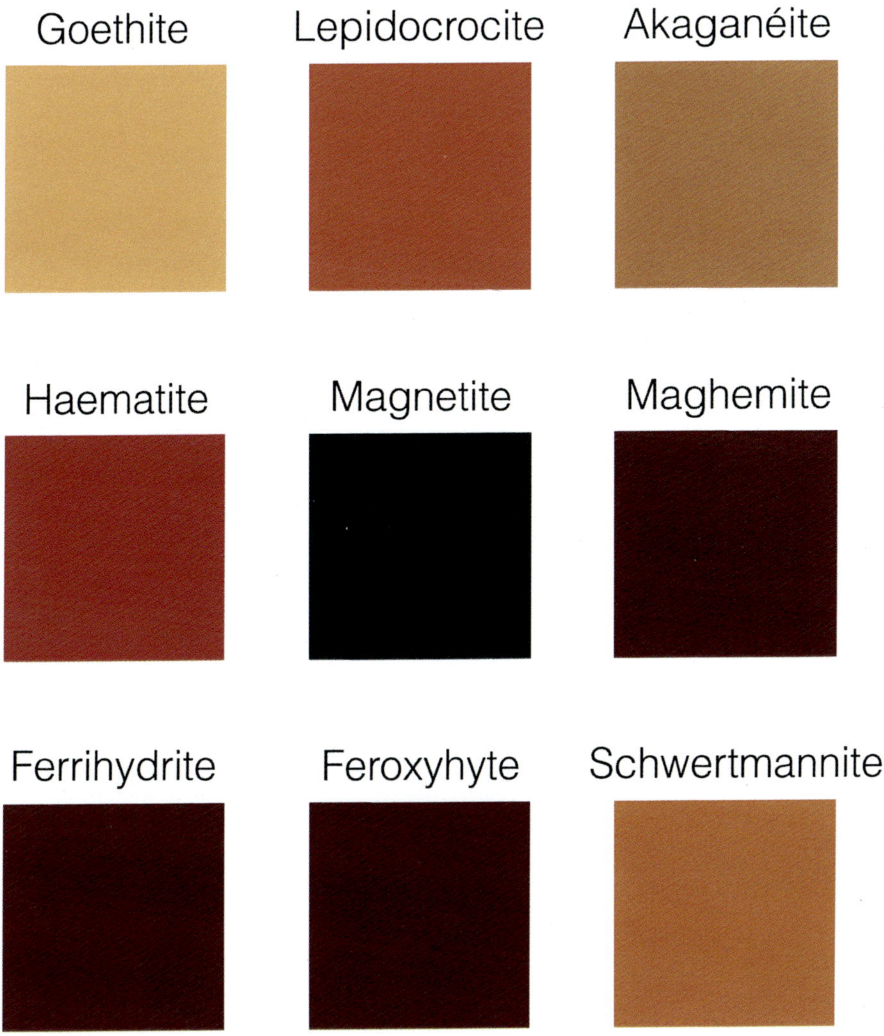

Plate V. Colors of iron oxides.

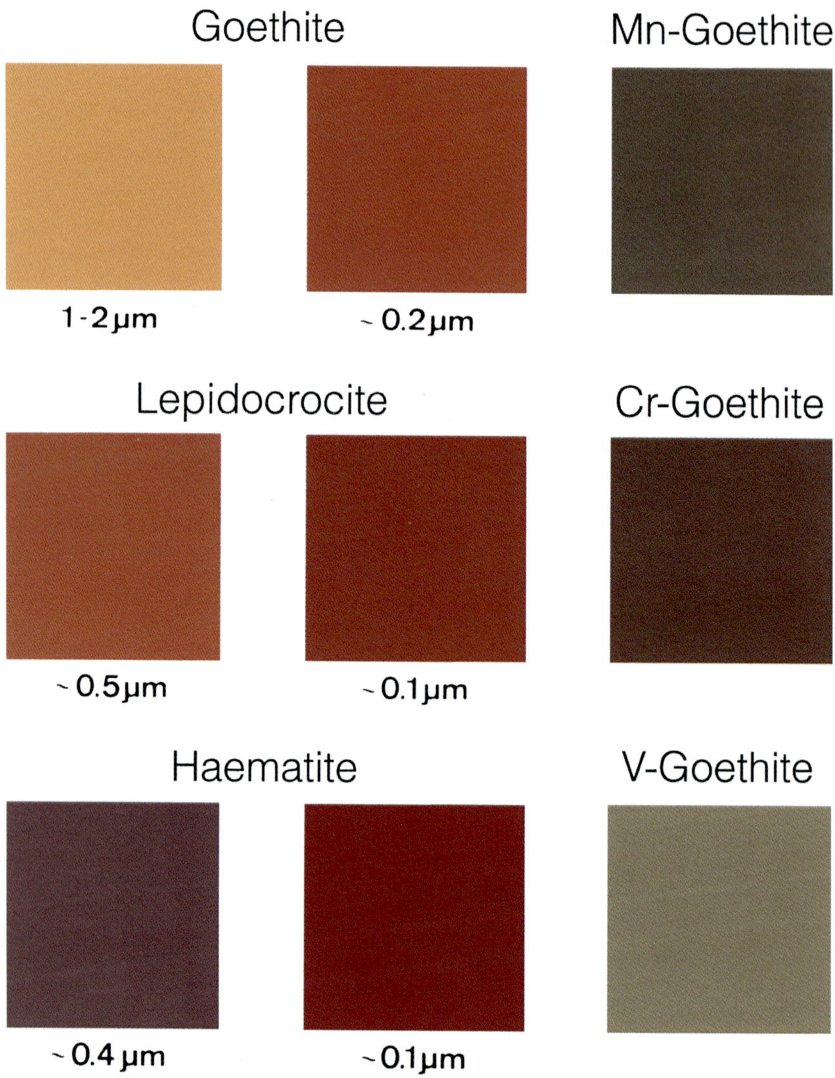

Plate VI. Effect of particle size and cation substitution on the color of iron oxides.

Plate VII. Self-supporting film of goethite (left) and hematite (right) (Cornell, unpublished).

Plate VIII. Microbiological formation of schwertmannite in a bioreactor with *Thiobacillus ferrooxidans* (Courtesy J.M. Bigham; Ohio University).

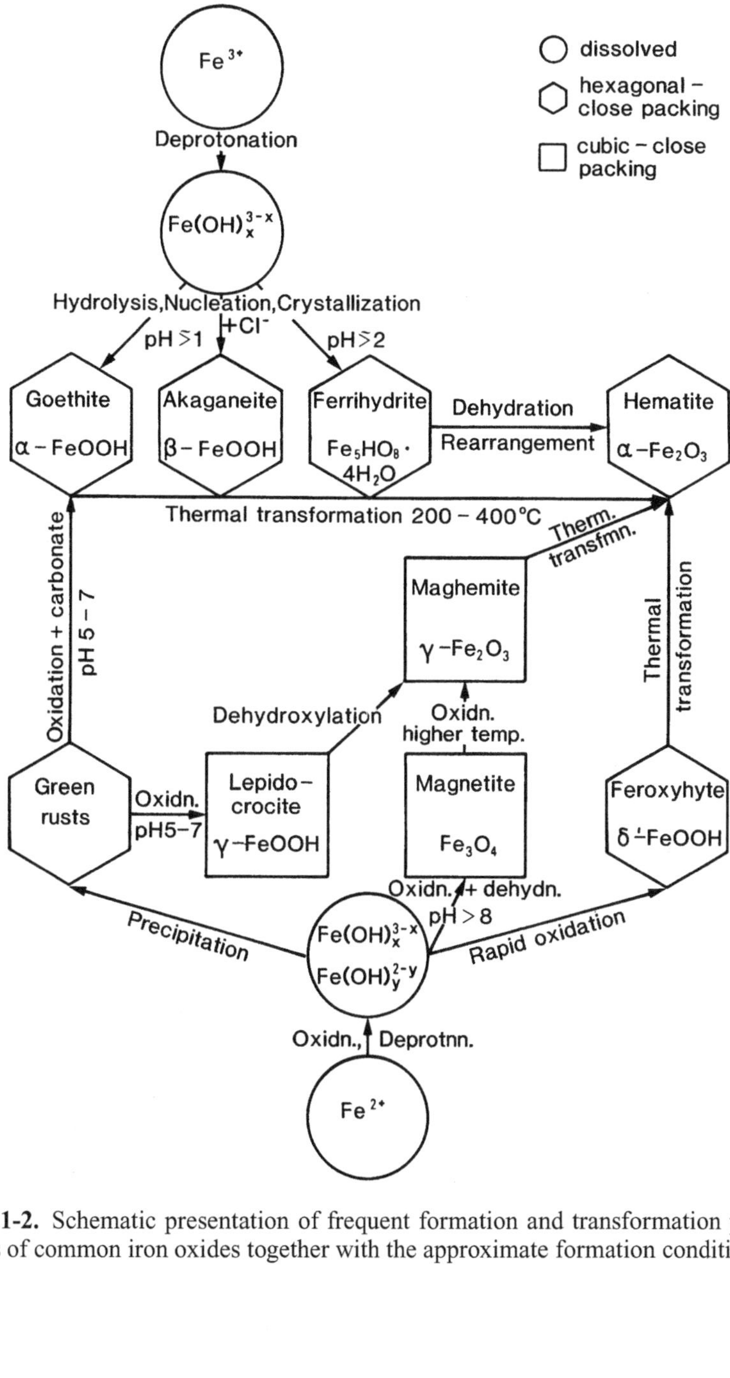

Fig. 1-2. Schematic presentation of frequent formation and transformation pathways of common iron oxides together with the approximate formation conditions.

Table 1-3. General characteristics of iron oxide minerals

Mineral name: Formula	Hematite α-Fe_2O_3	Maghemite γ-Fe_2O_3	Magnetite Fe_3O_4	Goethite* α-FeOOH	Lepidocrocite γ-FeOOH	Ferrihydrite variable	Feroxyhite δ'-FeOOH	Akaganeite β-FeOOH	Schwertmannite $Fe_8O_8(OH)_6SO_4$
Crystal system	Trigonal	Cubic or tetragonal	Cubic	Ortho rhombic	Ortho rhombic	Trigonal	Hexagonal	Momoclinic $\beta = 90.24°$	Tetragonal
Cell dimensions (nm)	a = 0.50340 c = 1.3752	a = 0.834	a = 0.839	a = 0.9956 b = 0.30215 c = 0.4608	a = 1.2520 b = 0.3873 c = 0.3071	a = 0.508 c = 0.94	a = 0.293 c = 0.460	a = 1.060 b = 0.3039 c = 1.0513	a = 1.066 c = 0.604
Formula units, per unit cell, Z	6	8	8	4	4	4	2	8	2
Density ($g\ m^{-2}$)	5.26	4.87	5.18	4.26	4.09	3.96	4.20	3.56	n. k.
Octahedral occupancy	2/3	–	–	1/2	1/2	< 2/3	1/2	1/2	1/2
Maximal Al-for-Fe substitution	1/6	+[l]	+[l]	1/3	+[l]	n.k.	n. k.	n. k.	n. k.
Standard free energy of formation $\Delta G°$ ($kJ\ mol^{-1}$)	–742.8[e]	–711.14[h]	–1016.1[i]	–488.6[e]	–477.7[j]	–699.0[h]	n. k.	–752.7[k]	n. k.
Solubility product (pFe + 3 pOH)	42.2–43.3	40.5	n. k.	43.3–44	40.6–42.5	37–39.4	n. k.	n. k.	n. k.
Type of magnetism	weakly ferromag or antiferromag.	ferrimag.	ferrimag.	antiferromag.	antiferomag.	canted antiferromag.	ferrimag.	antiferromag.	n. k.
Neel temperature (K)	955	n.k.	850	400	77	25–115[d]	440	250–300	75[d]

Points of zero charge (pzc) are generally between pH 8 and 9.

a Towe & Bradley (1967).
b Chukhrov et al. (1973)
c Russell (1979).
d Blocking temperature (Murad et al., 1988).
e Robie & Waldbaum (1967).
f Garrels & Christ (1965).
g Mohr et al. (1972
h Langmuir, (1969, 1971)
i Helgeson, 1969
j Van Schuylenborgh, 1973
k calc. from Murray, 1979
l Al-substitution possible but maximum not known
n. k. = not known.

* Goethite has long been placed in space group Pbnm. It has recently been placed in space group Pnma, necessitating a rotation of the crystal axes. In the new space group, a=0.9956, b=0.30 215, and c=0.4608 nm.Thus, what was previously the c-axis is now the b-axis.

β-Fe_2O_3 has been obtained by dehydration of β-FeOOH in high vacumn at 170 °C (Braun and Gallagher, 1972). ε-Fe_2O_3 can be produced by the reaction of alkaline potassium ferricyanide solution with sodium hypochlorite. It is also obtained (together with a mixture of other iron oxides) in an electric arc under an oxidizing atmosphere (Büttner, 1961). Its magnetic and thermal properties have been investigated by Dezsi and Coey (1973).

Wüstite, FeO, is obtained by heating a pelletized mixture of hematite and iron at 837 °C in a sealed silica tube for 24 hrs and then quenching in liquid N_2 (Battle and Cheetham, 1979). It is usually non-stoichiometric and contains defect clusters approaching the Fe_3O_4 structure. Wüstite is stable only at temperatures greater than 570 °C. At lower temperatures it decomposes to Fe_3O_4 and Fe.

1.3 Iron Oxides in the Environment

Iron oxides are widespread in nature. They are ubiquitous in soils and rocks, lakes and rivers, on the sea floor, in air (e. g. admixed in aeolian Sahara dust) and in organisms, and they may be responsible for the reddish-colored surface of the Mars. Iron oxides are of great significance for many of the properties and processes taking place in ecosystems.

In soils and sediments (e. g. in aquifers) iron oxides regulate the concentration of plant nutrients such as phosphate and inorganic (heavy metals) and organic pollutants and function as an adsorbent and an electron acceptor during dissimilatory metabolism of micro-organisms under anoxic conditions. In living organisms, Fe oxides may be present as an Fe reservoir such as ferritin (consisting of ferrihydrite), as hardening agents in teeth (goethite) or as a directional device in micro-organisms (magnetite). Where iron metabolic overload occurs, Fe oxides may form unwanted deposits in the human body.

Iron oxides may be either beneficial or undesirable. Everyone is familiar with rust, a mixture of various Fe oxides, as the end product of corrosion of iron. The modern steel industry, on the other hand, relies on the huge deposits of hematite and magnetite (iron ores) found in many parts of the world.

Iron oxides and hydroxides are introduced into the environment (pedosphere, hydrosphere, biosphere) from the lithosphere (earth's crust) during rock weathering. Iron is the fourth most abundant element (5.1 mass%) of the lithosphere. In primary (magmatic) rocks most of the Fe is located in Fe^{II} silicates such as pyroxenes, amphiboles, biotites and olivines; it can also be found in sulphides such as pyrite (FeS_2), and carbonates such as siderite ($FeCO_3$). During weathering at the surface, i.e. in the presence of water and oxygen, the silicates are decomposed by oxidation and hydrolysis:

$$-Fe^{II}-O-SiO_3 \xrightarrow{+O_2,+H_2O} -Fe^{III}-OH + H_4SiO_4 \tag{1}$$

For example, the formation of goethite (α-FeOOH) from an olivine (fayalite) (eq. 2) or from pyrite (eq. 3) can be written as

$$Fe_2SiO_4 + ½\,O_2 + 3\,H_2O \longrightarrow 2\,\alpha\text{-}FeOOH + H_4SiO_4 \tag{2}$$

$$4\,FeS_2 + 15\,O_2 + 10\,H_2O \longrightarrow 4\,\alpha\text{-}FeOOH + 8\,H_2SO_4 \tag{3}$$

In these reactions oxygen serves as the electron acceptor.

The resultant Fe^{III}oxides have a very low solubility (see Tab. 1-3), hence measurable concentrations of Fe^{3+} ions are only present under strongly acid conditions, e.g. those resulting from pyrite oxidation (eq. 3).

Once in the weathered zone (sediments, soils), the Fe of the Fe^{III}oxides may be remobilized under anaerobic conditions by microbial reduction:

$$4\,FeOOH + CH_2O + 8\,H^+ \longrightarrow 4\,Fe^{2+} + CO_2 + 7\,H_2O \tag{4}$$

This process involves enzymatic transfer of electrons from the biomass (written as CH_2O) to Fe^{III} during metabolic oxidation (dissimilatory respiration).

The iron oxides in natural surface environments are often poorly crystalline. i.e. the crystals are nano-sized (>100 nm), do not clearly exhibit the typical morphology of well-crystalline forms, are rich in defects and contain impurities. All this is most probably the result of their formation at low-temperature and in contaminated environments. Due to their striking colors (ranging from red to yellow) and their high surface area, small concentrations of Fe oxides color the soils and rocks in which they are mixed.

Precipitation, dissolution and reprecipitation of the various Fe oxides in the environment depend predominately on factors such as pH, Eh, temperature and water activity. For this reason, the different Fe oxides may serve as indicators of the type of environment in which they formed. Goethite and hematite are thermodynamically the most stable Fe oxides under aerobic surface conditions and they are, therefore, the most widespread Fe oxides in soils and sediments. Other Fe oxides are, however, also found in the environment because, although they are thermodynamically less stable, their formation is kinetically favored and their transformation to more stable forms proceeds sluggishly.

The yellow-brown *goethite* occurs in almost all soils and other surface formations (e. g. lakes, streams), whereas the red *hematite* often colors soils of tropical and subtropical regions. Higher temperatures and lower water activities in the latter regions are important parameters which partly account for this phenomenon. The ratio of goethite to hematite varies greatly with local conditions and is, therefore, an environmental indicator. The same is true for the extent of Al for Fe substitution; this ranges from 0 to 33 mol% for goethite and from 0–16 mol% for hematite. Its extent reflects, among other things, the availability of Al during Fe oxide formation. Goethite has also been found in the teeth of certain molluscs (limpets, chitons), but as yet, hematite has not been found in living organisms.

It is now generally believed that goethite precipitates directly in solution via a nucleation-crystal growth process, whereas formation of hematite requires the presence of ferrihydrite as a precursor; crystallization of hematite occurs within the ferrihydrite aggregates (Schwertmann et al. 1999). No direct, solid-state transformation of goethite to hematite by simple dehydration has so far been observed under *surface* conditions; it may, however, take place after sediment burial. Although soil goethites may be acicular or fibrous and hematites may appear as hexagonal plates, both oxides are most frequently isometric and irregularly shaped.

Lepidocrocite is generally less widespread than its polymorph, goethite, but it does occur frequently as orange accumulations in certain environments. These are characterized by the presence of Fe^{2+} from which lepidocrocite forms by oxidation. The presence of this mineral, therefore, indicates a deficiency of oxygen; in soils this is due mainly to excessive moisture. Lepidocrocite may exist on a pedogenic time scale (10^3 years) because, although it is metastable with respect to goethite, its formation may be kinetically favored and, furthermore, its transformation to goethite is extremely slow. In carbonate-rich solutions, formation of lepidocrocite is prevented and goethite forms (from Fe^{2+}) instead. Soil lepidocrocites resemble the lathlike crystals produced in the laboratory. Lepidocrocite is found in various biota inlcuding sponge spicules and the teeth of chitons.

Ferrihydrite occurs mainly in situations where Fe^{2+} is oxidized rapidly (in comparison with lepidocrocite formation) and/or where crystallization inhibitors are present. Oxidation can proceed via an inorganic pathway, but may also be assisted by micro-organisms such as *Gallionella* and *Lepthotrix*. Biological oxidation (termed *iron respiration*) supplies the micro-organisms with energy. Where micro-organisms have been involved, sheaths of bacteria filled with ferrihydrite particles may be found. Because of the high rate of oxidation during its formation and/or the presence of inhibitors, ferrihydrite is poorly crystalline with a very small particle size and hence a surface area greater than 200 m^2/g. Crystallization inhibitors include organics, phosphate and silicate species, all of which are widespread in natural environments and have a high affinity for the Fe oxide surface. The inhibitors also stabilize ferrihydrite and retard its transformation to more stable minerals.

Typical environments in which ferrihydrite exists are Fe containing springs (Plate II), drainage lines, lake oxide precipitates, ground water and stagnant-water soils and podsols, river sediments and, in the oceans, deep sea crusts and Mn nodules. Ferrihydrite is related to ferritin, an iron oxyhydroxide-phosphate association which acts as an iron reservoir in living organisms. The micelles of ferritin are encapsulated in a shell of protein molecules which prevents conversion to more crystalline, less active iron oxides thus maintaining the ability of the core to supply Fe.

In nature, ferrihydrite will slowly transform to hematite or goethite. Such a mechanism is feasible in view of the much higher solubility (lower stability) of ferrihydrite ($pK_s = 37–39$) in comparison with that of goethite and hematite ($pK_s = 40–44$). The transformation to goethite takes place over a wide pH range by a dissolution/reprecipitation process (see Fig. 1-2). It is slow at neutral pH (months to years), but can be greatly accelerated (days) by organic reducing agents such as cysteine. Ferrihydrite may, therefore, be considered to be a young Fe oxide in natural environments. This explains its occurrence in lakes, streams and hydromorphic soils, i.e. environments where oxidizing and reducing conditions alternate and hence an active iron turnover exists.

Maghemite is among those Fe oxides found invariably in the soils of the tropics and subtropics, but which also occur occasionally in soils of the temperate region. Maghemite may form during pedogenesis (soil formation) by several pathways. One is by oxidation of magnetite inherited from the parent rock or formed in soils. If derived from a titano-magnetite, maghemite may contain titanium in the structure. Another, probably widespread mechanism involves conversion of other Fe oxides such as goethite; the essential prerequisites are heat (from bush or forest fires) and the presence of organic matter. This mechanism may operate in temperate regions, but it is more common in the tropics because of the high frequency of fires. A further synthesis route which is often used in the laboratory, but is unlikely to operate in nature, involves dehydration of lepidocrocite. Maghemite formed by oxidation of magnetite is well ordered and may display superstructure lines (see Sec. 12.1), whereas firing of other Fe oxides produces a poorly ordered maghemite without superstructure lines and often containing structural Al.

Magnetite in soils is commonly inherited from the rock. It can, however, be formed in surface environments by biological processes and has been detected in various biota including bacteria (magnetotactic bacteria), bees and pigeons. The presence of magnetite is considered to be related to the directional sense of these organisms. Magnetotactic bateria have been identified in marine, limnic and soil (pedo-)environments (Fassbinder et al.; 1990). In contrast to the comparatively large size of magnetite crystals in rocks, the magnetite crystals in bacteria are only in the 0.1 μm range.

Akaganeite is formed in nature under the same conditions used in laboratory synthesis, i.e. in the presence of chloride and at elevated temperatures (e.g. 60 °C). Examples are the hot brines of the Atlantis Deep of the Red Sea and the hot springs of similar composition of the White Island volcano, New Zealand. Natural and synthetic akaganeites frequently display similar crystal morphologies.

Only a few natural occurrences of *feroxyhyte* have been reported so far; these were in some soils, in ochreous bands of pleistocene sediments and in marine concretions. Nothing is known about the mechanism of formation in nature, although rapid oxidation may be presumed to be involved.

Schwertmannite turned out be a common ochreous precipitate in water courses draining pyrite (FeS_2)-containing rocks and more widespread in so-called acid mine waters (Plate III). It is an oxidation product of pyrite in contact with atmospheric oxygen. The Fe^{2+} in these sulfate containing, extremely acid (pH <3) waters can only be oxidized with the assistance of a micro-organism, *Thiobacillus ferroxydans*. Formation of the Fe^{III}oxyhydroxy sulfate requires the presence of high SO_4^{2-} concentrations (ca. 1 g L^{-1}); associated iron minerals are goethite and jarosite, $KFe_3(SO_4)_2(OH)_6$. *Green Rusts* commonly form as a corrosion product of steel as long as the oxidation of the Fe is incomplete. Recently green rust has also been identified in poorly aerated, i.e. anaerobic soils (Trolard et al. 1997).

Further information concerning Fe oxides in natural environments is found in the reviews of Fitzpatrick (1988); Schwertmann (1985, 1988); Schwertmann and Taylor (1989), Taylor (1987) and Cornell and Schwertmann, (1996). Recent reviews of Fe oxide formation by biota are those of Loewenstamm and Kirschvink (1985), Mann et al. (1989) and Skinner and Fitzpatrick (1992).

2 General Preparative Techniques

2.1 Quantity of Product

The preparative procedures described in chapter 4 strictly hold only for the quantities (5–10 g) of product indicated. Simply scaling the quantities up or down to produce different amounts of iron oxide does not always produce the desired result. Crucial factors involved in altering the scale of a preparation are the ratio of reactant to Fe (e.g. OH/Fe or oxidant:Fe), suspension concentration, pH and temperature. Unless proved to be irrelevant, these variables should be held constant when the procedure is scaled up or down. Normally, this can be achieved quite readily when small reaction volumes (<0.5 L) are used. With larger volumes or when a reactant has to be added constantly (e.g. oxygen or base to a Fe^{II} system) this may be more difficult. The alternative then is to produce larger quantities in separate batches.

2.2 Treatment after Synthesis

2.2.1 Washing

If produced in an aqueous system, iron oxides must be washed after preparation in order to remove impurity ions associated with the procedure, e.g. OH^-, Cl^-, NO_3^-, SO_4^{2-}, CO_3^{2-}, NH_4^+, K^+, Na^+. After washing, the supernatant liquid should be tested for the presence of these ions. Washing may be carried out by a) centrifugation/decantation, b) vacuum filtration or c) dialysis.

a) *Centrifugation/decantation:* The product should be washed several times with twice distilled water until the electrolytes are removed. This method is rapid, but can lead to loss of product particularly the smaller particles. High sedimentation speeds (10 000 × g) or longer centrifugation times (1–2 hr) are needed to ensure complete sedimentation of particles smaller than ca. 0.1 μm. As the electrolyte is removed, the suspended particles tend to become increasingly dispersed, hence much longer centrifugation times (or higher speeds) are required to achieve sedimentation. Difficulties with highly dispersed suspensions may be overcome by adjusting the pH of the wash water close to the point of zero charge (PZC) of the oxide (usually pH 7–8) because at the PZC, flocculation is facilitated.

b) *Vacuum filtration:* filtration should be carried out with a 0.45 μm millipore filter under vacuum. With small quantities of liquid and solid and with solids with fairly large particles (> 0.1 μm) this method is fairly rapid and satisfactory. With large quantities of suspension or with very small particles (e. g. ferrihydrite) clogging of the filter makes the method impracticably slow.

c) *Dialysis:* this is probably the most effective method and involves holding the suspension in dialysis tubing in a vessel containing 1–2 l twice-distilled water. The water must be replaced at intervals and the necessary frequency of replacing the water can be determined with the help of conductivity measurements (using a conductivity meter). When the conductivity of the surrounding water equals that of distilled water, all the impurity ions have diffused out of the suspension. Dialysis takes between 1–2 weeks. The advantage of this method is that there is no loss of sample. Before use, the dialysis tubing* should be cleaned by boiling for two ten minute periods in twice distilled water.

Even after washing, the oxides may not be entirely free from adsorbed impurities. This applies particularly to anions such as sulphate and carbonate, which have a high affinity for the oxide surface. Even chloride and nitrate may be difficult to remove completely. A sensitive check for most of the adsorbed anionic impurities is provided by IR spectroscopy (see p. 51).

---

* Possible source: MEDICELL Int. Ltd. 239 Liverpool Rd. London N1 1LX

2.2.2 Drying

Oven drying at 40 °C is the most convenient method. This takes around 48 hours with 10 g sample (held in a wide mouthed vessel). A temperature of 40 °C is low enough to avoid any phase modification or transformation, except possibly in the cases of ferrihydrite and magnetite. After drying, the sample should be ground in an agate mortar to help break up any aggregates (see, however, p. 108 for ferrihydrite). Where difficulty in redispersion in water is expected (particularly with very fine grained material) freeze-drying is preferable to air-drying, but may not be entirely successful, e. g. for ferrihydrite. Freeze drying produces a loose dry powder which usually needs no grinding.

2.2.3 Storage

Iron oxides are most conveniently stored as dry powders. However, after prolonged storage in an air-dry state some metastable forms may transform into more stable ones. For example, ferrihydrite will gradually turn into hematite and goethite when kept in contact with the atmosphere, presumably owing to the presence of adsorbed non-stoichiometric water; Fig. 2-1 shows an X-

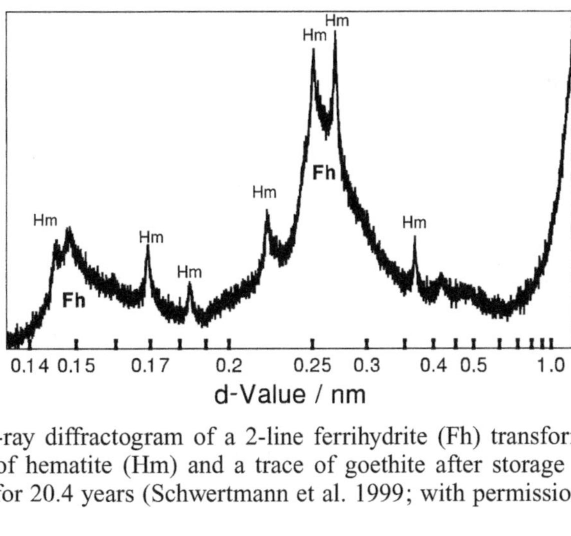

Fig. 2-1. X-ray diffractogram of a 2-line ferrihydrite (Fh) transformed to small percentage of hematite (Hm) and a trace of goethite after storage in an air-dry state at RT for 20.4 years (Schwertmann et al. 1999; with permission).

ray diffractogram of a 2-line ferrihydrite stored for 20.4 years in a glass vessel at room temperature (RT): hematite and a trace of goethite were formed.

Storing the iron oxides as suspensions has advantages and disadvantages. The advantage is that the difficulty of redispersion after drying is avoided; reasonably well crystalline goethite and hematite, have proved to be very stable in suspension for long periods of time (years). The disadvantages are: 1) unstable phases such as ferrihydrite will convert to more stable ones, 2) the particles tend to aggregate with time, and 3) bacterial growth may develop. The latter problem can be reduced by storing in a refrigerator (4 °C).

2.3 Chemical Analysis

2.3.1 Total Analysis

Total analysis consists of determining total Fe (Fe_t), Fe^{II}, adsorbed water, structural OH and, for substituted Fe oxides, such metals as Al, Mn, Cr and others.

Chemical analysis for Fe and other cations requires total dissolution of the oxides. This is usually achieved by treating the oxides (tenths to hundreds of mg) in conc. HCl (e.g. 0.1 g in 10–20 ml of 6–12 M HCl) at an elevated temperature (60–80 °C). The time needed for complete dissolution is normally less than one hour, but may be longer for crystals >0.5 µm. It may also be considerably longer for Cr-substituted Fe oxides (Schwertmann et al., 1989).

Atomic absorption spectroscopy is commonly used to determine Fe, Al, Mn, Cr and other metals. Standard solutions should be prepared with the same acid concentration as that of the test solutions. Apart from Al which requires a nitrous oxide/acetylene flame, these cations may be measured using an air/acetylene flame. These metals may also be measured by inductive coupled plasma analysis (ICP).

A non-destructive method of measuring Fe^{II} (e.g. in magnetite) is Mössbauer spectroscopy. Alternatively, Fe^{II} may be measured after dissolution of the oxide in M H_2SO_4 at 80 °C, using the α, α′-bipyridyl, the 1-10-o-phenanthroline or the ferrozine method.

All fine grained materials contain up to several percent of physically and chemically adsorbed water. The total water content of the FeOOH forms including structural OH can be determined from the weight difference found between samples heated at 105 °C and at 800 °C. The heating temperature which separates non-structural (chemically adsorbed) from structural water depends on the type of oxide and its crystallinity. For goethite thermal gravimetric analysis indicated that the temperature is between 160 and 180 °C (Schwertmann et al., 1985). This method applies only to well-defined FeOOH forms such as goethite. Hematites produced at low temperatures (< 100 °C) also contain a few percent water and OH, the latter being partly incorporated into the structure, possibly at Fe-deficient sites or other defects ("hydrohematite", Wolska, 1981). This water is driven off over a wide temperature range extending up to 800 °C.

2.3.2 Extent of Isomorphous Substitution

When other metals (M) substitute for Fe in the structure of an Fe oxide the mole ratio of substitution is given by $M_t/(M_t + Fe_t)$(mol/mol), where M_t and Fe_t (t = total) are expressed in mol. Fe and other metals present at the surface of the iron oxide or in separate phases must be determined separately to correct the extent of substitution. Ferrihydrite as a separate phase can be selectively dissolved an with acid oxalate solution (see p. 50). This treatment also dissolves any separate Mn or Cr oxides. Alternatively, a short extraction (30 min, 25 °C) with 0.4 M HCl removes adsorbed surface species; this method is useful if the solubility of the substituting ion in acid oxalate solution is not known or if the iron oxide under consideration (for example magnetite) is soluble in acid oxalate solution. The total Fe_t and M_t have then to be corrected for the oxalate soluble Fe and M.

2.3.3 Impurities

Chloride may be detected by precipitation as AgCl: dissolve the oxide in HNO_3, place a few ml solution in a test tube, and add one drop dilute

(0.1 M) $AgNO_3$ solution. The white precipitate of AgCl that forms in the presence of traces of Cl^- gives rise to a slightly turbid solution: the precipitate is more readily visible if the test tube is held against a black background. The limit of detection is ca. 1 μg Cl^-.

Nitrate may be detected with the use of diphenylamine: about 0.5 ml of a strongly acidic (sulphuric acid) diphenylamine solution is placed on a spot plate and a drop of test solution is placed in the middle; if NO_3^- is present, a blue ring (due to formation of the blue quinoid imonium ion) forms as the two liquids mix. The limit of detection is ca. 0.5 μg NO_3^-.

Sulfate may be detected by precipitation of barium sulphate. Dilute barium chloride solution (0.1 M) is added to the acid test solution in a test tube. If sulphate is present, the solution becomes turbid owing to precipitation of white $BaSO_4$. The limit of detection is ca. 5 μg SO_4^{2-}.

Ammonium can be recognised by the yellow precipitate that forms in the presence of Nessler's reagent. This is commercially available. It can also be prepared as follows: dissolve 14.3 g NaOH in 70 ml twice distilled water. Dissolve 5 g red HgI_2 and 4 g KI in 20 ml H_2O. Pour the iodide solution into the hydroxide solution and dilute to 100 ml. Allow to settle and use the clear supernatant liquid. To test for the presence of NH_4^+, the test solution is mixed with concentrated (1 M) KOH in a 1:1 ratio and a drop of the resulting suspension placed on a filter paper. Upon addition of a drop of Nessler's reagent a yellow orange precipitate of $HgI_2 \cdot HgNH_2I$ forms. The limit of detection is ca. 0.3 g NH_4^+.

Various methods exist for the quantitative analysis of these ions, the most convenient one being ion chromatography. For this purpose Fe should be removed prior to analysis by passing the acid solution through a strongly acid cation exchange resin followed by neutralization of the solution with the alkaline eluent (Bigham et al., 1990).

2.4 Removal of Iron Oxides from Reaction Vessels

The reaction vessels in which the iron oxides have been prepared are usually stained with the product. The oxide hydroxides and hematite may be removed by overnight treatment (25 °C) with 4 M HCl (or in difficult cases with 37% HCl). Magnetite may be removed in 1–2 hours by treatment with 0.5 M oxalic acid at 25 °C or with M H_2SO_4 at 80 °C.

2.5 Purity of Reagents

All reagents should be Analar grade. In addition, care must be taken that the solid Fe^{II} or Fe^{III} salt is perfectly fresh and has not been exposed to the atmosphere for long periods. Iron salts can undergo complicated reactions. Some hydrated Fe^{III} salts may dehydrate. $FeCl_3 \cdot 6\,H_2O$ on the other hand is hygroscopic, i. e. it takes up water upon exposure to the atmosphere and in the salt solution which forms on the surface, hydrolysis may take place leading to an ochreous coating of akaganeite. This precipitate can be removed from an Fe^{II} solution by filtration. Fe^{II} salts may oxidize as well as dehydrate. Color changes indicate whether such reactions have taken place; for example $Fe(NO_3)_3 \cdot 9\,H_2O$ develops orange flecks among the bulk purple crystals and greenish $FeSO_4 \cdot 7\,H_2O$ and $FeCl_2 \cdot 4\,H_2O$ develop an orange color. Such discolored reagents must not be used.

Fe^{III} and Fe^{II} solutions should be used immediately after preparation. If Fe^{III} solutions are allowed to stand, some nucleation of goethite can take place, even at room temperature. If Al-substituted goethite is being prepared, for example, the presence of such seed crystals of goethite in an aged Fe^{II} solution will prevent a uniformly substituted product from being obtained. Aged Fe^{II} solutions will, of course, be partly oxidized.

3 Methods of Characterization

3.1 Introduction

Because the synthetic iron oxides whose synthesis is described in this book consist of very small crystals ranging in size from ca. 3 nm to several μm, they are characterized and checked for purity by the techniques commonly used for submicroscopic to nano-sized particles. The most important of these techniques are X-ray powder diffraction and electron microscopy. In addition, Mössbauer spectroscopy, infra red absorption spectroscopy and thermal analysis provide useful information (Wilson, 1987).

This chapter provides a brief account of the potential of these methods to characterize the iron oxides described here. The methods were also used in principally the same way to produce the analytical information given for each oxide under "Product description". For details of the theory and application of these techniques, the reader is referred to the relevant treatises that are cited, where appropiate, in the text. The major diagnostic criteria for the common Fe oxides obtained with these methods are assembled in Tab. 3-1.

3.2 Color (A. Scheinost)

One of the most striking features of iron oxides is their eye-catching color (see Plates V and VI). The colors include purple, red, brown, orange, yellow, black, and even greenish blue (green rust). To some extent, these colors are characteristic and, thus, diagnostic of the type of mineral, its crystal size and shape, and impurities within the crystal

Table 3-1. Diagnostic criteria of iron oxide minerals

Mineral	Color (Munsell)	Usual crystal shape	Most intense XRD spacings, nm	DTA events, °C	IR bands, cm^{-1}	Magnetic hyperfine field (T) 295 K	4.2 K
Hematite	red 5R – 2.5YR	Hexagonal prisms, Rhombohedra	0.270, 0.368, 0.252	Nil	345, 470, 540	51.75	54.2/53.5[a]
Maghemite	red to brown	Cubes	0.252, 0.295	Ex[d] 600–800	400, 450, 570, 590, 630	45–52	50.2/51.7[a]
Magnetite	black	Cubes, octahedra	0.253, 0.297	see footnote[b]	400, 590	49.1/46.0[c]	–
Goethite	brownish-reddish yellow 7.5YR – 10YR	Needles, laths	0.418, 0.245, 0.269	En[d] 280–400	890, 797	38.2	50.6
Lepidocrocite	reddish yellow 5YR – 7.5YR	Laths	0.626, 0.329, 0.247, 0.1937	En 300–350 Ex 370–500	1026, 1161, 753	–	45.8
Akaganeite	brownish yellow 10YR	Somatoids, rods	0.333, 0.255, 0.7467	Ex ca. 400	1050, 820, 670, 410		47.3, 47.8, 48,9
Ferrihydrite	dark reddish brown 5YR – 7.5YR	Spheres	0.254, 0.224, 0.197, 0.173, 0.147	En 150–200 Ex 300–350		–	49–49.5
Feroxyhyte	dark reddish brown 5YR – 7.5YR	Fibers, Needles	0.254, 0.222, 0.169, 0.147	En 250	1110, 920, 790, 670	–	53.0/50.8
Schwertmannite	reddish yellow 7–9YR	Hedgehogs	0.255, 0.339, 0.486, 0.151	En 200; 700 Ex 560	1186, 1124, 1038, 976, 704, 608, 483	–	45.4/45.8

a With and without Morin transition, respectively.
b Magnetite converts via maghemite or directly to hematite, depending on particle size.
c For tetrahedral and octahedral Fe, respectively
d Ex: exotherm; En: endotherm

structure. Even small differences in color can be effectively distinguished with the naked eye, and easily quantified by means of color charts (e.g., the Munsell color book) or color meters. It is this ease and speed of determination which makes color a helpful tool in mineral synthesis, even though the mineral identification by other methods like XRD, FTIR and Mössbauer spectroscopy is generally more reliable. In this section, we will give a short introduction into the physical causes of the colors of iron oxides, color measurement and color systems, followed by data tables and qualitative descriptions which will help to interpret the colors of mineral samples produced in the laboratory.

3.2.1 Origin of Color

Color is caused by the absorption of light in the visible range of the electromagnetic spectrum, i.e. between ca. 400 nm (red) and ca. 700 nm (purple). The corresponding photon energies are 1.8 to 3.1 eV. In this energy range, light may interact with the valence electrons of iron oxides in three different ways:

(1) Oxygen-to-metal charge transfer (OMCT) gives rise to an intense absorption band in the UV which is broad enough to extend a wing into the higher-energy (blue) side of the visible spectrum.

(2) Over this wing are superposed crystal-field bands due to electronic transitions within the incompletely filled 3d-shell of Fe^{III}. Owing to its high-spin $3d^5$ electron configuration and the octahedral or (more rarely) tetrahedral coordination in the iron oxides, these transitions would be expected to be forbidden by both spin and dipole (Laporte) selection rules and hence only weakly absorb light. An example of this situation is the pale purple color of Fe^{III} nitrate. However, in iron oxides the intensities of these crystal-field bands are strongly enhanced (Fig. 3-1). The reason for this is that the Fe^{III} centers in edge- or face-sharing octahedra are close enough to each other to cause magnetic spin coupling of the 3d electrons. Furthermore, spectroscopic selection rules for coupled Fe^{III}-Fe^{III} pairs differ from those for isolated Fe^{III} centers, giving rise to additional transitions by simultaneous excitation within two adjacent Fe^{III}, the so-called electron pair transition (EPT in Fig. 3-1).

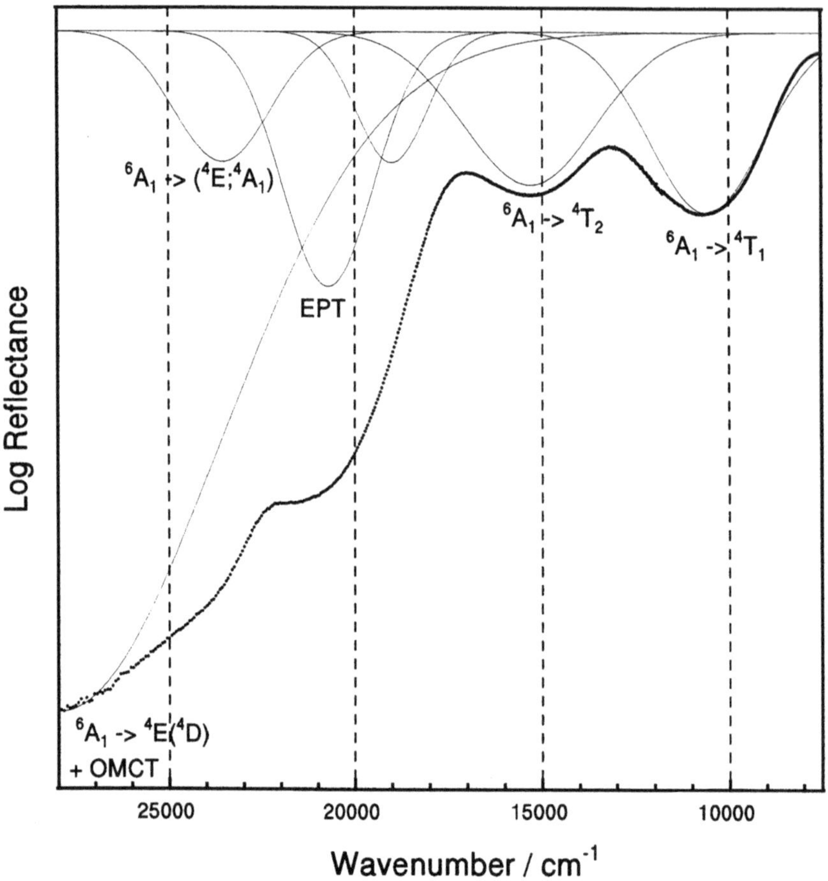

Fig. 3-1. Band assignment for the diffuse reflectance spectrum of a goethite. Note that one band could not be assigned. EPT: electron pair transition; OMCT: oxygen-metal charge transfer. From Scheinost et al., 1999; with permission.

(3) Intervalence charge transfer (IVCT) may occur between neighboring Fe^{II} and Fe^{III} centers, leading to spin-allowed absorption bands. In the case of green rust, such an IVCT strongly absorbs in the red portion of the visible spectrum, giving rise to the greenish color. The IVCT of magnetite centers at about 1500 nm, but extends across almost all of the visible spectrum, hence magnetite is black and has a metallic luster.

The colors of iron oxides are caused by at least two of these electron transitions (1 and 2) as illustrated for the spectrum of goethite (Fig. 3-1).

Absorption of one part of the visible spectrum causes selective reflection of the remaining (complementary) part. This reflected light is what is detected by the eye (or any color measuring equipment) and translated into a color sensation by the brain. For example, a pigment absorbing light between 400 and 500 nm (blue to green) reflects between 500 and 700 nm (yellow to red), and hence appears yellow to orange (see spectrum of goethite in Fig. 3-2). As the absorption edge extends further into yellow, the pigment appears red (see spectrum of hematite in Fig. 3-2). Hence, a relatively small change in a mineral's structure causing a slight shift of the absorption edge, may rather drastically affect its hue. Furthermore, the slope and profile of the absorption edge also modifies color. Generally, narrow absorption bands causing steep absorption edges produce more vibrant colors than do broad bands.

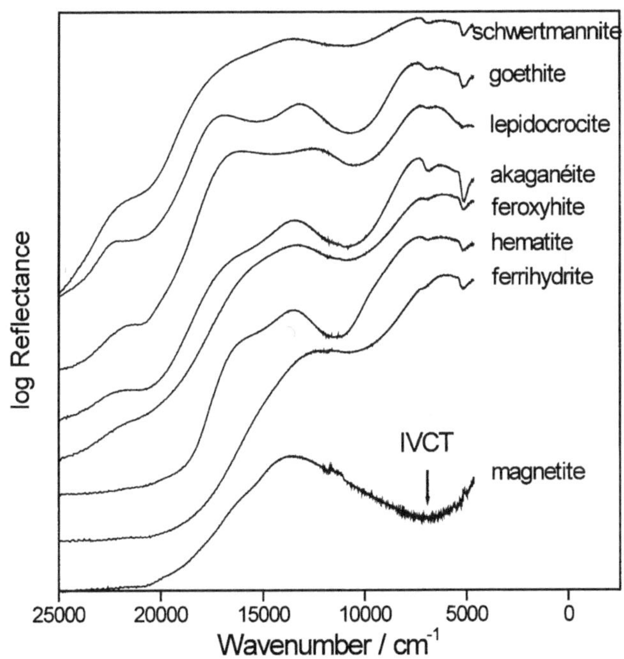

Fig. 3-2. Typical diffuse reflectance spectra of Fe oxides. Courtesy A. C. Scheinost.

When the particle size of a pigment is close to the wavelength of visible light, its color may be further altered by elastic scattering of the incident and the reflected light. Thus, particle size and shape, as well as the refractive indices of the pigment and embedding medium, may further modify the color which is primarily caused by the various absorption processes described above. The color of a pigment is also altered by diluting it with a non-coloring (white) agent, which influences the wavelength-dependent absorption and scattering. The color of high pigment dilutions may be predicted by applying an equation developed by Kubelka and Munk (1931).

More detailed information on the physical causes of color can be obtained from books by Nassau (1983) and Burns (1993).

3.2.2 Color Measurement

Color Charts. The colors of pigments can be routinely measured by visual comparison with color charts like the one issued by the Munsell Color Company. This method may achieve a high precision because of the eye's sensitivity to small (relative) color changes. Of paramount importance for reliable application is a uniform light source. The major drawback of the method is that the resolution of color assignments is limited by the resolution of the color chips (see 3.2.3). Interpolation between chips is possible in principle, but difficult in practice as it has to be done in a three-dimensional space.

Color Meter. The Chroma Meter CR-300 (Minolta, Osaka, Japan) provides both accurate and precise color data (Post et al., 1993). This color meter uses a pulsed xenon light for stable and uniform illumination of the sample. Three photocells measure the photons diffusely reflected by the sample through filters matching the CIE standard observer spectral response. A standard white plate is used for calibration. For analysis, the hand-held detector unit which is protected by a glass window may be placed on top of a small pile of a powdered sample (300 mg), thus standardizing the sample surface. This way, up to one hundred samples can be precisely measured within one hour. Further advantages are the mobility of the unit, and the output of the three color parameters in several color systems (see below).

Diffuse Reflectance Spectrophotometer. This instrument measures the wavelength-dependent reflectance of a surface relative to a white standard. The color can be calculated from these spectra by means of computer programs (Fernandez and Schulze, 1987). In addition, absorption band positions may be extracted from the spectra by using mathematical procedures to enhance the resolution of the generally broad and strongly overlapping bands (Fig. 3-2). These methods include Fourier self-deconvolution (Kauppinen et al., 1981), derivation (Huguenin and Jones, 1986), and band fits with Gaussian distributions (Sunshine et al., 1990). The latter approach has been further refined by constraining band positions according to ligand field theory (Scheinost et al., 1998). The bands shown in Fig. 3-1 have been fitted in this way. Standard transmission spectrophotometers covering the visible spectrum can be converted into diffuse reflectance spectrometers by using appropriate attachments (either hollow spheres coated on the inside with Halon, a white reflecting material, or so-called praying-mantis attachments). Recently, the speed of collecting a spectrum has been greatly enhanced by employment of diode array detectors.

Sample preparation. Due to the influence of particle size and the particle-embedding medium on light scattering, both parameters have to be standardized. A good method for producing isolated particles in air is shock-freezing of a particle suspension followed by freeze-drying. Aggregated particles may be gently broken up and ground until no further change in color is observed (Torrent and Barrón, 1993). However, the color of iron oxides may also be measured in suspension, e. g. during synthesis in the mother liquor, or immersed in oil.

3.2.3 Color Systems

A color can be numerically defined as a point in a three-dimensional color space. In the past, a vast number of color spaces, optimized for different applications have been developed. The most physical of these is the CIE-Yxy space, where all colors lie within a plane resembling a shoe sole. An arbitrary point B within that plane is defined by the Cartesian coordinates x and y, where the redness increases with x, and the greenness increases with y. The lightness, Y, is perpendicular to x and y out of the plane. An alternative way of defining B is by drawing a line through B and the achromaticity point, A. The intersection of that centripetal line with the perimeter gives the dominant wavelength, λ_d, (580 nm for point B in Fig. 3-3). The relative distance between A and the intersection with the perimeter gives the excitation purity, P_e (0.6 for point B). The dominant wavelength is therefore similar to the hue, and P_e is similar to the chroma of the Munsell system.

Equal Euclidean distances in the CIE-Yxy system are not necessarily perceived by a human observer as equal color distances. To match Euclidean and perceived distances, a wide variety of so-called uniform color spaces has been developed (Hunter and Harold, 1987). One of the first of these uniform systems was developed in the early 1900's by the artist A.H. Munsell. Starting with the observation that brighter hues like yellow and red create a much stronger color sensation at color saturation than darker hues like blue or green, he mounted color chips on a rotating Maxwell disk to

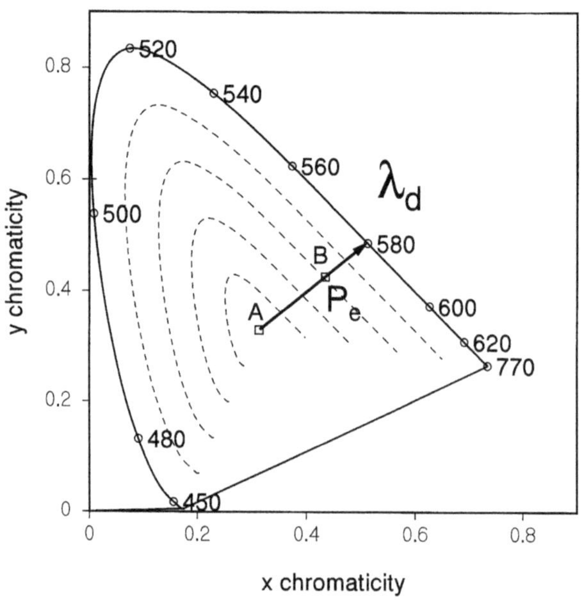

Fig. 3-3. Color designation in the CIE-Yxy system. Colors lie within the solid lines resembling a shoe sole. Colors within that plane may be given either by the Cartesian coordinates x and y, or by the dominant wavelength, λ_d [nm], and the excitation purity, P_e. The lightness, Y, is normal to the plane. (Scheinost and Schwertmann; 1999; with permission)

find pairs of equal intensity perception (chroma), which he subsequently arranged in a three-dimensional manner, often referred to as color tree (Plate IV). An improved version of Munsell's color tree soon became an industry standard. Nowadays, the Munsell color book contains about 1500 color chips. A color point in this system is defined by the three parameters-hue, value and chroma. Hue refers to the spectral colors red (R), yellow (Y), green (G), blue (B), purple (P) and their intermediates, (e.g., YR or RP). These 10 major hues are arranged in a circle of 100° and further subdivided in 1° steps (e.g., 10 YR or 5 Y, Fig. 3-4). With increasing distance from the achromatic center of this circle, N, the color intensity, or chroma, increases from 0 in steps of one. The third parameter, value, defines the lightness, and extends perpendicular to the hue-chroma plane, with its greatest extension in N reaching from 0 (black) over shades of gray to 10 (white). With increasing chroma, the range in value decreases. Due to the above mentioned unequal perception of bright versus dark hues, the resulting cylindrical color space is irregularly filled.

In contrast to the patchwork-like Munsell system, the CIE system and its modifications (developed by the International Commission on Illumination, CIE, in 1931) create numerically continuous color spaces. They are based on the principle of trichromaticity of vision which states that the

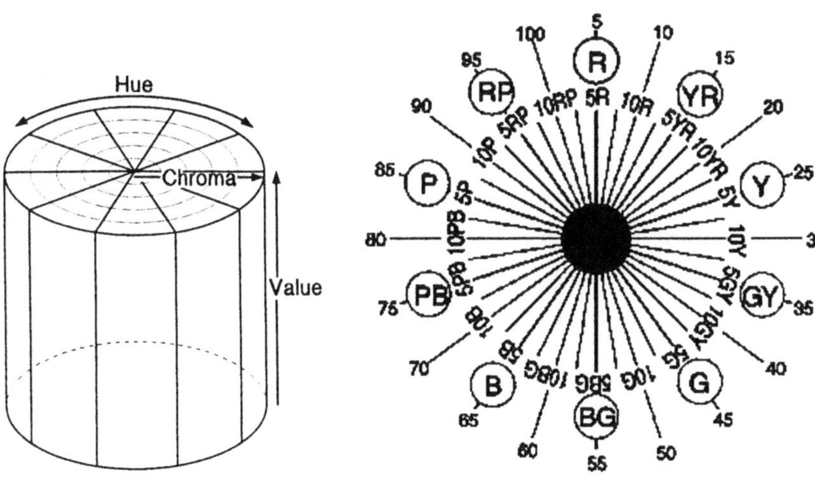

Fig. 3-4. The cylindrical coordinates of the Munsell color systems (left) and the hue notation (right). (Figure provided by Munsell, a division of GretagMacbeth, LLC.)

human eye responds to only three color components, corresponding to the tristimulus values X, Y, and Z. The tristimulus values can be mathematically converted to create uniform color spaces, one of them being the CIE-Lab system with the Cartesian coordinates L* (lightness, corresponding to value), a* (redness-greenness), and b* (yellowness-blueness, Fig. 3-5). Alternatively, the Cartesian coordinates a* and b* may be re-

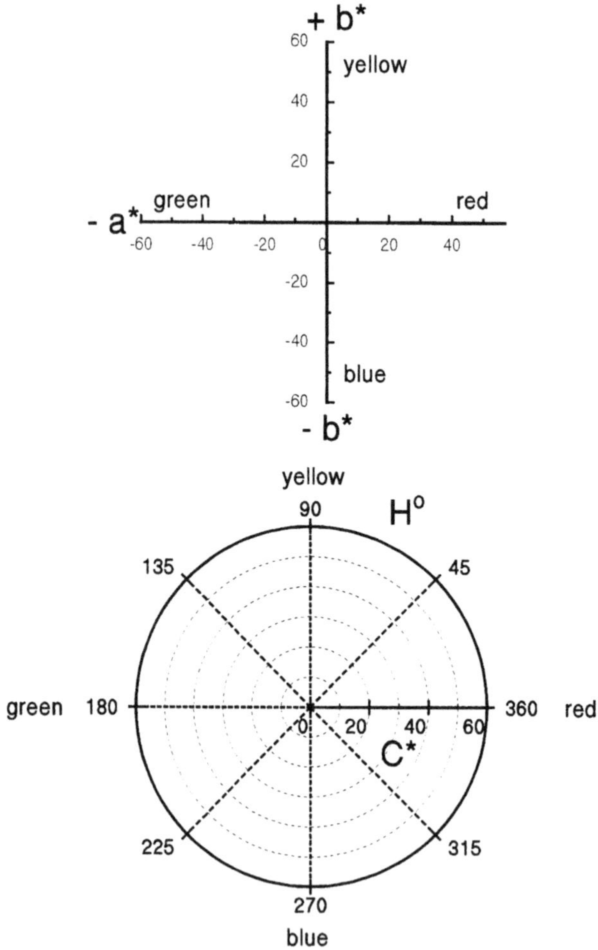

Fig. 3-5. The CIE-Lab color system in Cartesian notation (a*b*, left), and in polar notation (C*H°, right). (Scheinost and Schwertmann; 1999; with permission)

placed by the polar coordinates C* (CIE-chroma) and H° (CIE-hue). H° describes a circle of 360°, where 0° corresponds to red, 90° to yellow, 180° to green, and 270° to blue (Fig. 3-5). While Cartesian coordinates are the best choice for any statistical data treatment of color points, colors in cylindrical coordinates are easier to visualize.

Note that the uniformity of a color space is of concern only if color is regarded as a (visual) quality. When using color as an analytical tool, the color system which differentiates the iron oxides most effectively would be the most favorable, no matter whether it is uniform or not. A discriminant analysis performed with a large number of iron oxide samples showed that CIE-Yxy, CIE-Lab, and the Munsell system performed equally well (Scheinost and Schwertmann, 1999). Therefore, the choice of a color system is not crucial. Furthermore, computer programs are available which transfer colors from one system into another. More details on color theory and color systems are given by Wyszecki and Styles (1982) and Hunter and Harold (1987).

3.2.4 Identification of Iron Oxides by Color and Crystal-Field Bands

Tables 3-2 and 3-3 give the color ranges of 8 iron oxide minerals in the two most common color systems, Munsell and CIE-Lab. These tables have been compiled from Chroma Meter measurements of 165 synthetic samples. Hematite has the most reddish average hue, while feroxyhite, ferrihydrite, akaganéite, lepidocrocite, maghemite, schwertmannite and goethite exhibit increasingly yellower hues. The redder hues of hematite and feroxyhite are due to face-sharing octahedra, where adjacent Fe centers are as close together as 0.29 nm, thus strongly enhancing EPT. In contrast, the other minerals are only built up from edge- and corner-sharing octahedra, where Fe centers are 0.3–0.33 and 0.35 nm apart, respectively, giving rise to more yellowish hues. Magnetite and partly oxidized magnetite have the darkest values due to the existence of Fe^{II} and Fe^{III} causing IVCT bands. The average values of the other, purely Fe^{III} minerals rank similarly to their average hues. This ranking reflects the fact that red hues appear darker to the eye than yellow hues. The darker values of maghemite, hematite, feroxyhite, akaganéite and ferrihydrite correspond also to their lower chromas.

Table 3-2. Munsell colors of synthetic Fe oxide samples (mean, minimum – maximum)

	Hue	Value	Chroma	N
Akaganéite	5.4 YR 4.4 YR–8.0 YR	3.9 2.8–5.0	5.8 4.4–6.9	8
Ferrihydrite	4.4 YR 2.8 YR–5.5 YR	3.8 2.3–5.4	4.7 1.9–6.6	10
Feroxyhite	3.9 YR 1,5 YR–5.4 YR	3.8 3.4–4.7	5.9 4.2–7.0	11
Goethite	0.5 Y 8.1 YR–1.6 Y	6.1 4.9–6.9	7.1 6.2–9.6	54
Hematite	0.3 YR 3.5 R–4.1 YR	3.6 2.4–4.4	3.6 2.4–4.4	44
Lepidocrocite	6.7 YR 4.9 YR–7.4 YR	5.5 4.6–5.9	8.3 7.3–9.9	26
Maghemite	8.3 YR 6.2 YR–9.4 YR	3.1 2.5–3.6	3.2 2.5–4.1	7
Schwertmannite	8.8 YR 7.7 YR–9.2 YR	6.2 5.6–6.5	7.9 7.3–9.1	5

In spite of a substantial variability of color within the minerals, goethite, hematite, lepidocrocite and ferrihydrite can be easily distinguished by hue and chroma (Fig. 3-6). Hematite is always redder than 4.1 YR, and goethite is yellower than 8.1 YR. Ferrihydrite and lepidocrocite have intermediate hues, and are separated by chroma. The color of maghemite is unique due to values lower than 3.6 and hues yellower than 6.2 YR. Feroxyhite overlaps with hematite in hue and value, but exhibits a higher chroma than hematite. Akaganéite overlaps with ferrihydrite in hue, value and chroma. Only akaganéite samples with a hue yellower than 5.5 YR may be reliably distinguished from ferrihydrite. Finally, schwertmannite samples may be distinguished from goethite only if they are redder than 8.1 YR. In conclusion, for six out of eight iron oxide minerals color is a reliable indicator of a successful mineral synthesis*.

* Natural iron oxide samples, often less crystalline and less pure than samples made in the laboratory are more difficult to identify by color (Scheinost and Schwertmann, 1999). They may also be cemented into dense masses the colors of which are usually much darker than those of powders of the same material (Schwertmann, 1993)

Table 3-3. CIE L*C*H° colors of synthetic Fe oxide samples (mean, minimum – maximum)

	L*	C*	H°	N
Akaganéite	40 29–51	35 25–43	63 55–71	8
Ferrihydrite	39 23–56	27 10–39	57 44–63	10
Feroxyhite	39 35–48	34 24–41	56 45–63	11
Goethite	63 51–71	47 41–62	80 72–84	54
Hematite	37 25–45	29 9–42	42 21–57	44
Lepidocrocite	57 47–61	50 45–60	68 63–70	26
Maghemite	32 25–37	20 15–24	68 61–72	7
Schwertmannite	64 58–66	50 45–58	75 71–77	5

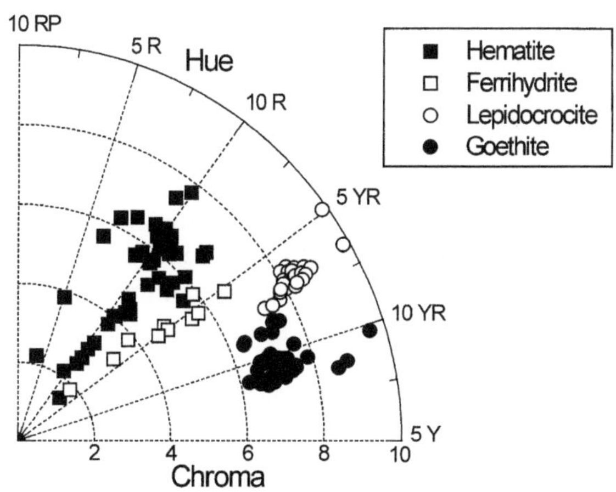

Fig. 3-6. Munsell colors of hematite, ferrihydrite, lepidocrocite, and goethite samples. (Figure provided by Munsell, a division of GretagMacbeth, LLC.)

Depending on the type of mineral, individual samples may deviate from the average by up to 3.8 units of hue (hematite), 1.6 units of value (ferrihydrite), and 2.8 units of chroma (ferrihydrite). This intra-mineral variability can be explained by differences in crystallinity, particle shape, particle size, and aggregation. Decreasing crystallinity may cause a broader distribution of electron transition bands, and hence smoother absorption spectra with a less vibrant appearance. The particle shape influences the orientation of particles towards the incident light. For instance, platy particles may be oriented along one crystallographic axis, while spherical particles may absorb along randomly oriented axes. As the absorption spectra vary with the crystallographic orientation, this may have a profound influence on the color. Thirdly, mean particle size and its distribution and the aggregation of particles influence the scattering of the incident and the reflected light, further modifying the appearance of the color.

Due to the large number of possible interactions, quantitative relationships with color have been established for a few cases only. The color of hematite changes from yellow-red to blue-red as the diameter of crystals increases from 0.1 to 1.0 µm (Schwertmann, 1993) (PlateVI). A similar effect is obtained by the aggregation of hematite crystals (Torrent and Schwertmann, 1987). Acicular hematite crystals have a more yellow hue than hexagonal hematite crystals (Hund, 1981). While well crystalline goethite appears moderately yellow, the color shifts towards darker values and redder hues as the needle length decreases from 0.8 to 0.05 µm (Scheinost et al., 1999) making the goethite look more brownish (see also chap. 5.2.3). In a similar manner, the color of lepidocrocite changes from the bright orange of larger crystals to the dark brownish-orange of smaller crystals (Plate VI).

Substituting part of the iron atoms by other metals also has an influence on the color. In the case of Al substitution, the observed shift toward redder hues is predominantly due to the correlated decrease in particle size (Scheinost et al., 1999). An increase in value (lightness) with increasing Al substitution was found for synthetic Al-hematites (Barron and Torrent, 1984). Other transition metals add their own electron transition bands to those of iron, leading to a red shift with increasing amounts of Cr, Co and Ni, and to a green shift with increasing V and Mn (Fig. 3-7).

Diffuse reflectance spectra (DRS) like those shown in Fig. 3-2 may also be used to distinguish between several iron oxide minerals. Band posi-

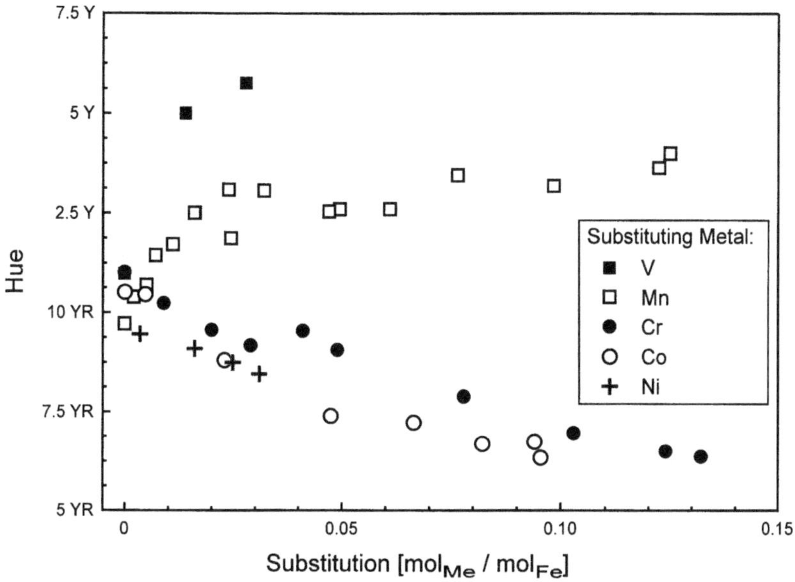

Fig. 3-7. Munsell hue of goethite as a function of substitution with transition metals. (Courtesy A. C. Scheinost, ETH2.)

tions can be determined using second derivatives of the measured spectra after applying a cubic-spline data smoothing (Scheinost and Schwertmann, 1997). While the reliability of mineral discrimination based on DRS is comparable to that based on color, the advantage of employing DRS is in detecting hematite, maghemite and magnetite impurities in other iron oxides. The EPT of hematite is at about 540 nm, hence at significantly higher wavelength than the EPT of other iron oxides (<500 nm, Fig. 3-8). The unique characteristic feature for magnetite and incompletely oxidized maghemite is the IVCT band at about 1500 nm (Fig. 3-2).

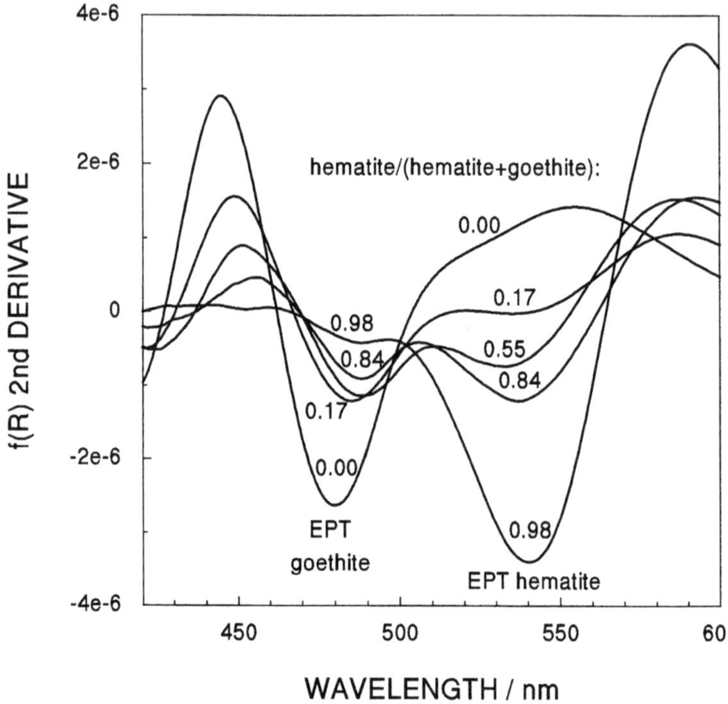

Fig. 3-8. Second-derivative spectra of hematite/goethite mixtures. With permission from Scheinost et al. (1998).

3.3 X-Ray Powder Diffraction

A mineral is defined by its structure, i.e. by the regular arrangement of its atoms in space. Only those methods, therefore, which reflect the structure are capable of providing unambiguous ***identification*** of a particular oxide. In general, diffraction methods fulfill this purpose and X-ray powder diffraction (XRD), the most common of these, is essential for identification and control of the purity of the product. The minimum size required for a crystal to diffract X-rays is of the order of a few unit cells (ca. 2–3 nm). Electron diffraction is another method often used for iron oxides. In addition to phase identification, XRD provides information about crystal size (and hence surface area) and crystal perfection, structural parameters (unit cell edge lengths) and degree of substitution of Fe by other trivalent cations (isomorphous substitution). A full account of the theory and practice of X-ray diffraction is provided in many textbooks (e.g. Klug & Alexander, 1974; Brindley & Brown, 1980; Bish & Post, 1989). In this section, a brief account of certain features which are relevant for the characterization of iron oxides, is presented.

The X-ray diffraction pattern of a powdered sample is a plot of the observed diffracted intensity of the X-rays against the Bragg angle, i.e. the angle at which the X-rays strike the crystal and for which the maximum interference is observed. An XRD patttern of a crystalline phase, therefore, consists of a number of reflections (peaks) of different intensities which provide a fingerprint of the atomic structure. From these patterns the mineral-specific distances between the atomic layers (d-values) can be calculated using the Bragg equation and from the set of d-values the mineral(s) can be identified. The strongest XRD peaks for the different Fe oxides are listed in Tab. 1-3. The full set of d-values for each mineral can be found in the Powder Diffraction File (International Centre for Diffraction Data; Newton Square, Pennsylvania 19073-3273, USA).

The X-ray pattern is preferably recorded stepwise using a goniometer equipped with a proportional or scintillation counter and a registration unit. The type of X-ray radiation normally used for Fe oxides is CoKα (as used in this book) or FeKα with a wave length of 0.178890 and 0.193604 nm, respectively. CuKα radiation is less suitable because it is strongly absorbed by Fe rich phases leading to loss of X-ray intensity and a high background due to fluorescence radiation. The latter can be removed with a monochromater.

How the reflected X-rays are collected with respect to angular range, step size and counting time per step will depend on the information required. For quick identification of an oxide and its possible impurities, a limited angular range comprising the strongest diffraction peaks (Tab. 3-1), wide angular steps (e.g. 0.05–0.1°2θ) and short counting times per step (1–5 s) are usually sufficient. To characterize the product in more detail, however, wider angular ranges (down to d-values of ca. 0.1 nm), smaller steps (e.g. 0.02°2θ) and longer counting times (10–20 s) are necessary. Poorly crystalline, i.e. weakly diffracting oxides, such as ferrihydrites, with broad peaks require longer counting times but tolerate larger steps.

To measure the angular position, intensity and shape of the peaks, the step-counted pattern is usually computer-fitted using either empirical single-peak models based on a range of different peak profiles or, more recently, the Rietveld model which produces a whole-pattern fit. The Rietveld model compares a calculated pattern based on the simultaneously refined model for crystal structure, diffraction optics, instrumental factors and other characteristics with the experimental pattern. Preferred orientation, asymmetric crystal development and peak broadening due to small crystals and structural disorder can also be accounted for. Numerous versions of the fit program exist and further improvements are continuously produced (see httd//sun7.unige.ch/stxnews/stx/welcome.htm).

To correct for instrumental misalignment an internal standard, e.g. 10% silicon (Si) or corundum (α-Al_2O_3) powder (1 µm; Fisher Comp. 122651 K) should be mixed with the sample. The peak width must be corrected for the instrumental broadening using a well crystalline standard, such as lanthanum boride LaB. As an example, a Rietveld fit of an X-ray pattern from a mixture of synthetic hematite and goethite is shown in Fig. 3-9.

If the peak positions are used to determine the unit cell edge lengths, a further correction of the peak position (so far not incorporated in available Rietveld models) is needed for very small crystals the broad peaks of which may show an (apparent) shift $\Delta°2\theta$ due to inconstancy of the Lorenz-polarization and structure factor over the angular range of the peak.

This correction increases as the full width of the peaks at half height (FWHH) becomes broader. Schulze (1984) calculated this shift for the four strongest reflections of goethite (Fig. 3-10 upper) and Stanjek (1991) for those of hematite (Fig. 3-10; lower). As seen from the two figures, a correction for both oxides is only necessary for a corrected FWHH greater than ca. 0.6°2θ which corresponds to a crystal size of ca. 10–20 nm. The figures show that the correction is very much hkl-dependent.

The corrected FWHHs can be used to further characterize the ***crystal size*** (more precisely the size of the mean coherently scattering domain, MCD) in the various crystallographic directions if this is smaller than ca.100 nm. For example, small hematite crystals are often less developed in the [001]

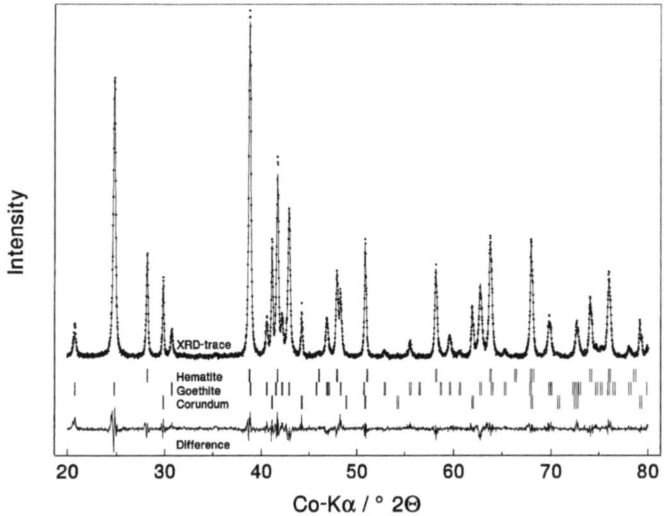

Fig. 3-9. Rietveld-fit of an X-ray diffractogram. The sample is an artificial mixture of synthetic hematite, goethite and corundum (internal standard) in a ratio of a:b:c. The X-ray diffractogram (upper) is followed by the x-ray line positions of the three components (middle) and the differences between the measured and the fitted intensities (residuals). Courtesy J. Friedl and H. Stanjek).

than in the [hk0] direction, because of their platy, i. e. anisotopic morphology, yielding broadened (hkl) and sharp (hk0) peaks. It must be taken into account, however, that peak broadening is also caused by structural disorder. The term ***crystallinity*** is therefore often used as a collective term under which both, crystal size *and* disorder are grouped together.

The degree of ***isomorphous substitution*** of Fe by other cations can be determined by chemical analysis provided that the sample consists only of one phase and that all of the foreign cations are incorporated into the structure. A definite proof of structural incorporation can, however, only be deduced from a shift in the position of the XRD peaks. A shift only occurs if the replacing cation is sufficiently different in size from that of the Fe^{III} ion. The most widely studied case for such a replacement is that of Fe^{III} (r = 0.064 nm) by the smaller Al (r = 0.053nm) (see Tab. 1-2). This leads to slightly smaller unit cell as indicated by a shift of all peaks towards higher angles. Changes in the unit cell size have also been measured for a range of di-, tri- and tetravalent metal cations including the

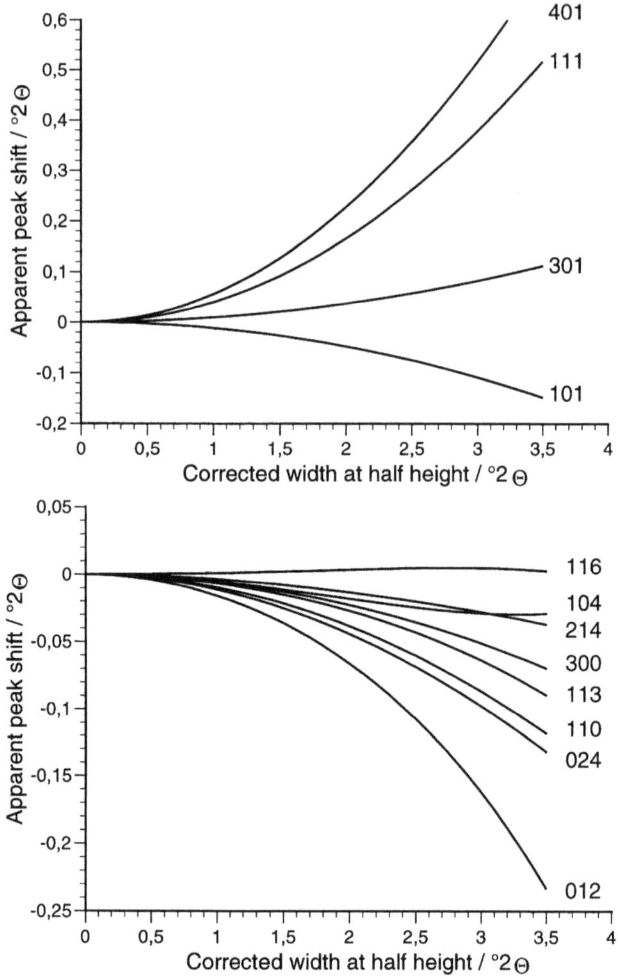

Fig. 3-10. Apparent shift of major XRD peaks as a function of corrected width at half height (FWHH) for goethite (upper) and hematite (lower). From Schulze, (1984); with permission and Stanjek, (1991); with permission.

potentially toxic metals Co, Ni, Zn, Cd, Pb. The extent of the peak shift can be used to quantify the extent of isomorphous substitution even in mixtures of several oxides. The accuracy for Al substitution is about 0.02 mol per mol of Fe.

Theoretically the unit cell edge lengths of a mixed phase (solid solution) should lie on a straight line between those of the two end members, e. g. goethite and diaspore, AlOOH, (Vegard rule, Vegard, 1921). For Fe^{III} oxides, however, the change in unit cell edge length is usually smaller than that expected from the Vegard rule. This can be due to incorporation of (additional) OH into the structure as shown for Al-substituted hematites synthesized at low temperatures. Heating of such hematites removes OH and thus lowers the unit cell size (Wolska, 1988, Stanjek & Schwertmann, 1992; Schwertmann et al. 2000). Similar effects can be observed for Al-goethites (Fey & Dixon, 1981; Schwertmann & Carlson, 1994; Schwertmann & Stanjek, 1998). In other words, different linear relationships are obtained for oxides produced in different ways. Furthermore, although Mn^{III} is of the same size as high-spin Fe^{III}, it causes the unit cell of goethite to expand in the *a* direction and shrink in the *b* and *c* directions owing to the so-called Jahn-Teller effect which distorts the coordination sphere.

In summary, because the relative change in cell edge lengths depends on the way in which the samples are synthesized, the degree of substitution of an unknown sample can only be obtained from relationships between the chemical composition and the crystallographic data for monomineralic phases which have been synthesized in the same or similar way.

A review of such relationships for various oxides and cations is given in Cornell & Schwertmann (1996).

3.4 Microscopy

Microscopy enables crystal size, morphology and domain character to be observed directly. Light microscopy is not suitable for the preparations described in this book, because the particles are too small. Electron microscopy enables much higher magnifications to be obtained; with High Resolution Electron Microscopy (HRTEM) the lattice fringes may be imaged. Scanning Electron Microscopy (SEM) produces three dimensional images at moderate magnification, whereas Transmission Electron Microscopy (TEM), provides two dimensional images of crystals ranging in size from a few nm to one or two μm across. The comparatively recently developed Scanning Probe Microscopy (Binnig et al.; 1986) permits three-dimensional observations at the nanometer or even atomic level. This technique operates in two different ways. With Scanning Tunneling Microscopy (STM) a sharp metallic tip scans closely (<1 nm) over a conducting or semiconducting surface. Its application to minerals has been reviewed by Hochella (1995) and Eggleston (1994). For Atomic Force Microscopy (AFM), on the other hand, the sample (which can be non-conducting) is mounted on a piezoelectric tube, rastered with a sharp tip mounted on a cantilever and the reflection of the cantilever monitored (Maurice, 1998). Applications of both techniques to Fe oxides are found in Hochella, (1995); Maurice et al. (1995); Weidler et al. (1996) and Forsythe et al. (1998).

Of all these techniques, TEM is the one most routinely used to examine the Fe oxides produced by the recipes in this book. For TEM examination a drop of a suspension of sample in twice distilled water (dispersed with ultrasonic treatment if necessary) is placed on a carbon coated, copper grid. For HRTEM, a holey carbon grid is used. A dilute suspension (0.01% by weight) is essential to avoid both a crowded grid and aggregation of particles which make crystal shape and size difficult to determine. The crystal lengths and widths are measured directly from the electron micrograph. The heights of the particles must be found by shadowing the sample with chromium at an angle of 45°. The length of the shadow cast is then equal to the crystal height. An example is shown in Fig. 3-11. From the lengths of the shadows it can be concluded that the crystals consist of flat laths. To obtain the average particle size and its frequency distribution, enough particles from a representative sample must be measured so that the average dimensions show no change when additional particles are included. This number depends on the uniformity of the particles. Schulze and Schwertmann (1984) counted between 60 and 90 particles on magnified TEM images to determine the dependance of the lengths and widths of acicular goethite crystals on Al-substitution.

Prolonged exposure of the sample to the electron beam may cause radiation damage particularly in the case of akaganeite (Galbraith et al.; 1979). A technique for imaging particles in aqueous suspensions in a supercooled, vitrous state is Cryogenic TEM. It has been applied to Fe oxides by Bailey et al. (1993).

Fig. 3-11. Lepidocrocite crystals shadowed with 5.0 nm chromium at 45°. The sample was synthesized by oxidizing an $FeCl_2$ solution with air at 50 °C. (Photo and recipe- R. Giovanoli, Univ. Bern).

3.5 Surface Area, Porosity and Fractal Dimensions

Essential characteristics of fine-grained solids, such as the Fe oxides, include the specific surface area, the porosity and the fractal structure. The standard procedure for measuring these properties is the Brunauer-Emmet-Teller (BET) method (Gregg & Sing, 1991) This depends on the fact that if a gas is brought in contact with a powdered solid (adsorbent) at a temperature near the condensation temperature of the gas, the gas molecules may form a monolayer at the surface from which the surface area can be calculated if the area occupied by one gas molecule is known.

Before adsorption, the solid must be evacuated overnight at a temperature of between 100 and 150 °C to remove physically adsorbed water which would otherwise interfere with the N_2 adsorption. Usually, this treatment does not lead to phase changes or transformations in most of the iron oxides. Although temperatures of between 90 and 120 °C are necessary to remove surface water from ferrihydrite, the phase transformation is blocked once adsorbed water is removed (Weidler, 1997).

The better crystalline iron oxides do not have an internal surface area, hence the BET method measures the total surface area of the solid. Ferrihydrite, however, frequently contains aggregates the internal surfaces of which may not be entirely accessible to N_2, hence the BET surface area is much lower than that calculated from the average particle size obtained by electron microscope observation. For example, for spherical particles with a diameter of 3 nm and a density of 3 g/cm^3 a 2-line ferrihydrite would have a surface area of ca. 600 m^2/g whereas its N_2-BET surface is hardly ever higher than ca 300 m^2/g. Indeed, ferrihydrites are usually strongly aggregated irrespective of whether dried in air or freeze-dried.

N_2 adsorption is also used to estimate the *micro pore volume and the pore size distribution* (see e.g. Glasauer et al. 1999) which can be derived from a plot of adsorbed N_2 vs. the thickness of a statistical monolayer, t, which is a function of the relative gas pressure (t-plot method). Mercury porosimetry serves the same purpose (Celis et al. 1998). N_2 adsorption isotherms have also been used to determine the *fractal dimensions* of Fe oxide particles (c.f.. Celis et al. 1998; Weidler et al. 1998).

3.6 Acid Oxalate Extraction

When an iron oxide such as goethite or hematite is produced from 2-line ferrihydrite, the product can be purified by selectively dissolving the remaining ferrihydrite. This method is also used to estimate the degree and rate of conversion of ferrihydrite.

In the absence of light, a buffered oxalate solution which consists of 0.2 M ammonium oxalate brought to pH 3 with 0.2 M oxalic acid, dissolves ferrihydrite at RT within 2 hours while leaving the crystalline phases essentially ($\lesssim 2\%$, depending on crystallinity; Schwertmann, 1973) undissolved (Schwertmann, 1959 a; 1964). The ratio of oxalate soluble Fe (Fe_o) to total Fe (Fe_t), Fe_o/Fe_t, (Fe_t is found by complete dissolution of the sample) gives the proportion of ferrihydrite in the sample. During the extraction the sample should be agitated. Light can be excluded by wrapping the reaction vessel in aluminium foil or using a dark room. In the presence of light goethite and hematite will dissolve as well as ferrihydrite (Cornell & Schindler, 1987). The amount of oxalate solution used should not be less than approximately 1 ml per mg oxalate-soluble Fe. If less oxalate solution is used, yellow Fe^{II}-oxalate may precipitate in the filtered extract and hence the amount of Fe_o in the sample will be underestimated. If oxalate extraction cannot be used because of oxalate adsorption by the undissolved part of the sample or because the sample contains magnetite, it can be replaced by a short extraction (30 min, 25 °C) with 0.4 M HCl.

How precisely the above methods will separate ferrihydrite from better crystalline oxides depends on the form and crystallinity of the latter. A positive relationship was found between Fe_o/Fe_t on the one hand and the surface area and XRD line width on the other for 14 synthetic goethites (range of $Fe_o/Fe_t = 0.003–0.05$) and 15 synthetic lepidocrocites (range of $Fe_o/Fe_t = 0.06–077$) (Schwertmann, 1973). This shows that whereas goethites (and hematites) are essentially insoluble in oxalate irrespective of their crystal size, the method can only be used for well crystalline lepidocrocites. The same applies to the use of dilute strong acids.

3.7 Infrared Spectroscopy

Infrared spectra arise as a result of interactions of solids with electromagnetic radiation (photons) in the wavelength range 1–500 µm (i. e. wave numbers of 10,000–20 cm^{-1}). These interactions involve excitation of vibrations or rotations of molecules in their electronic ground state and are associated with stretching deformations of the interatomic bonds and bending deformations of the interbond angles. The frequency of radiation absorbed depends upon the rotational energy levels and the force constants of the interatomic bonds (Farmer, 1974; Griffiths & de Haseth, 1986).

A transmission infrared spectrum is a plot of percent radiation absorbed versus the frequency of the incident radiation given in wavenumbers (cm^{-1}) or, occasionally, in wave lengths (μm). Increased resolution is provided by Fourier Transform IR (FTIR) which averages a large number of spectra and gives an improved signal to noise ratio and hence increased sensitivity compared with conventional IR spectroscopy. FTIR also permits more rapid collection of data. A variation of this method is reflectance spectroscopy: this may be external (Diffuse reflectance-DR FTIR) or internal (attenuated total reflectance-ATR FTIR). DR-FTIR is used for samples with poor transmittance, e.g. cubic hematite crystals. ATR-FTIR is suitable for investigation of suspensions of oxides and has been used to study the surface chemistry of goethite and interactions with adsorbing species (Tejedor-Tejedor & Anderson, 1986).

For Fe oxides, FTIR spectroscopy provides a rapid means of identification. It can detect traces (1–2%) of goethite in a sample of hematite. In addition, low levels of impurities arising from insufficient washing, for example nitrate in ferrihydrite (band at 1384 cm^{-1}) and carbonate in goethite (ca. 1300 and 1500 cm^{-1}) can be detected. Broadening of the absorption bands reflects a decrease in crystal perfection (Cambier, 1986). Structural substitution of Fe by Al (Schulze & Schwertmann, 1984), Mn (Stiers & Schwertmann 1985), and Cr (Schwertmann et al. 1989) causes a shift in the positions of the bands.

The most convenient method of sample preparation for routine identification of Fe oxides is DR-FTIR: just load the powder in a holder. Thin self-supporting films of goethite may also be prepared for IR examination by vacuum filtering a goethite suspension (0.1 g/50 ml water) through a 0.45 μm filter paper (Cornell, 1983). Such films are useful for gas adsorption studies and for examination of surface hydroxyl groups. They can be converted into self-supporting hematite films by heating at 350 °C (see Plate VII).

3.8 Thermoanalysis

A sample is continuously heated at a constant rate (e.g. 10 °C min^{-1}) while two changes are recorded: (1) the temperature difference between an inert compound and the sample with a thermocouple (differential thermo analysis; DTA) and (2) the weight loss measured with a balance (thermogravimetry; TGA) (Mackenzie, 1957; Smykatz-Kloss, 1974). With DTA, information is obtained about endothermic and exothermic phase transformations (see Fig. 1-2), whereas with TGA adsorbed water and structural OH can be measured.

All fine grained Fe oxides lose adsorbed water at characteristic temperatures of between 100 and 200 °C. Structural OH in goethite and lepidocrocite is lost at 250–400 °C by the dehydroxylation reaction: $2\,OH \rightarrow O + H_2O$. Even fine grained oxides such as hematite contain some OH in the structure (Stanjek & Schwertmann, 1992) and this is driven off over a wide temperature range. For Fe oxides endothermic peaks result from the release of adsorbed or structural water, whereas exothermic peaks come from phase transformations (e.g. maghemite to hematite) or from recrystallization of smaller crystals into larger ones. An example of this is observed during the transformation of ferrihydrite to hematite.

3.9 Mössbauer Spectroscopy

Among the spectroscopic methods, Mössbauer spectroscopy (nuclear resonant γ-ray absorption spectroscopy) has the great advantage of being element-specific and as ^{57}Fe is a convenient Mössbauer active isotope, the technique is very suitable for iron oxides. It supplies information about the magnetic field at the nucleus, the valence of the Fe and the type of coordination and order within the ligand shell. Common magnetic hyperfine properties extracted from Mössbauer spectra are the isomer shift, the quadrupole splitting and the magnetic hyperfine field.

When the oxide particles are very small (tens of nm), fluctuations of the electron spin direction may be so fast that the nucleus can no longer follow them and therefore senses a net zero magnetic field. This phenomenon is called superparamagnetism. The spin reorientation can be slowed down by reducing the sample temperature eventually as far down as 4 K and the temperature at which the nucleus again senses a magnetic field can be used to calculate the particle size. Valuable additional information about the magnetic ordering behavior of small crystals can be obtained from the temperature dependence of the hyperfine parameters. Al as a diamagnetic ion reduces the magnetic ordering temperature if incorporated into the structure.

For reviews on the subject dealing with Fe oxides the reader is referred to Murad and Johnston, (1987) and Chapter 7.5 in Cornell & Schwertmann, (1996).

4 Synthesis Pathways

4.1 Nucleation and Crystal Growth

Formation of iron oxides in aqueous systems involves nucleation and crystal growth. A brief summary of these topics is provided as a background to the synthesis methods described in this chapter. Fuller details can be obtained in the books by Nielson (1964), Walton (1967) and Mullin (1993).

4.1.1 Nucleation

Homogeneous nucleation occurs spontaneously in bulk solution when the supersaturation* exceeds a certain critical value. The essential requirement for precipitation to take place is the formation of stable, embryonic clusters of molecules or ions, the so-called nuclei, in solution. These embryonic nuclei form as the result of collisions of ions or molecules in the bulk solution prior to nucleation. The embryos are continually breaking up and reforming. The stability of an embryo depends upon a balance being achieved between the free energy necessary for the creation of a new interface, the interfacial or surface energy, $\Delta G_{surface}$, and the energy released by the formation of bonds in the bulk structure, ΔG_{bulk}. The free energy of nucleation, ΔG_N is the sum of these two energies. Only when the embryo exceeds a certain critical size, does ΔG_{bulk} predominate. From this point on, the nucleus grows by a decrease of free energy. The critical nucleus, i. e. the smallest embryo that can actually

* A solution is supersaturated with respect to a certain compound if the ion product of this compound exceeds its solubility product.

continue to grow to a crystal becomes smaller, the lower the interfacial free energy.

The crucial parameters governing ΔG_N are the interfacial energy of the nucleus, $\Delta G_{surface}$, and the supersaturation of the solution. As the rate of nucleation, J_N, is related to ΔG_N, these parameters also influence J_N. The rate of nucleation is slow until a critical supersaturation is achieved, after which it rises rapidly.

When a solid can exist as two or more phases with different solubilities (e.g. as ferrihydrite and as goethite), the initial precipitate is often the more soluble phase, although it might be expected that the less soluble phase should form first because, for this phase, the supersaturation is higher. However, as the rate of nucleation depends on interfacial energy as well as on the degree of supersaturation, the more soluble phase may precipitate first, if the interfacial energy of the critical nucleus of this phase is lower than that of the critical nucleus of the less soluble phase. Owing to its higher solubility, the initial precipitate may, however, subsequently transform into the less soluble phase. This behavior, whereby the unstable polymorph forms first, or, in other words, kinetic factors override thermodynamics, is referred to as the Ostwald Law of Stages: it occurs quite often in the iron oxide system.

Heterogeneous nucleation occurs when the presence of a solid phase reduces ΔG_N and thus leads to an increase in J_N. The substrate may be minute particles of a foreign phase or crystals (seeds) of the phase that is crystallizing.

Heterogeneous nucleation can take place at a lower level of supersaturation than is required for homogeneous nucleation. The suitability of a foreign solid as a substrate is mainly a question of the degree of matching between structure type and atomic distances, rather than the chemical similarity of the two solids. With a good match between the structures of the crystal and the substrate ($<20\%$ difference) the interfacial free energy difference between a given crystal face and the substrate is lower than that between the same crystal face and the solution, hence nucleation is facilitated.

Examples of both homogeneous and heterogeneous nucleation exist in the iron oxide system. Goethite precipitates directly from soluble ferric species in solution. Its formation can, however, be assisted by addition of seed crystals of goethite to the system. As the interplanar spacings in the structures of goethite and hematite are comparable, hematite crystals can also serve as a substrate for heterogeneous nucleation of goethite giving rise to epitaxial* twins (Fig. 5-14). In solid-state transformations, nuclei form at sites at which local ordering has occurred (e.g. for hematite formation from ferrihydrite) or in regions of increased structural strain (e.g. during dehydration of lepidocrocite to maghemite). In such cases, addition of seed crystals of the product to the system does not promote further nucleation. If local ordering within the solid precursor can be induced, however, nucleation of the product can be promoted. In the case of hematite this is achieved by adsorption of oxalate (a chelating ligand) on ferrihydrite (Fischer and Schwertmann, 1975).

4.1.2 Crystal Growth

Although a certain degree of supersaturation is needed for crystal growth, it is much less than that required for nucleation. Crystal growth involves a number of steps including diffusion of the growth units to the crystal surface and diffusion and adsorption processes at the surface itself. The overall rate of growth is determined by the slowest of these steps.

The habit of the crystal is governed by the rates of growth of the different faces. Those faces which grow slowly tend to persist, whereas fast growing faces are eliminated. Foreign species which are adsorbed on the surface of the growing crystal may alter the rates at which the different faces grow and thus cause a change in crystal habit. This is a common phenomenon in the iron oxide system.

4.1.3 Production of Monodispersed Particles

Iron oxide samples often display a wide range of particle sizes indicating that nucleation and crystal growth took place simultaneously over the bulk of the reaction. In recent years, attention has been focussed on the production of oxides with different shapes and also with a narrow range of particle sizes, so-called monodispersed products (see Fig. 10-6). This

* Epitaxy: Growth of one phase on the surface of a crystal of another phase.

interest has been stimulated by the fact that magnetic, optical, electronic and catalytic properties can vary markedly with particle size and also shape. Monodispersed Fe oxides form ideal systems for fundamental investigations into these properties. They are also often used for investigating adsorption and dissolution processes.

The mechanisms by which monodisperse particles form are not yet fully understood although a large number of papers, especially on monodispersed hematite with various crystal shapes, has been published by, for example, Matijevic and coworkers (Matijevic and Partch, 2000) and by Sugimoto and coworkers (cf section 10.4). Originally, it was considered that the necessary requirements for production of monodisperse particles are separation of nucleation and crystal growth and a continuous supply of growth units, i. e. reactive, low-molecular weight ionic species. The separation of nucleation and growth can be achieved by arranging the reaction conditions so that there is slow generation of growth units until the critical supersaturation for nucleation is exceeded, at which point the supersaturation is reduced by a burst of nucleation. Thereafter, the growth units are taken up by the nuclei. Their rate of formation must be sufficiently slow so that they are removed entirely by the nuclei i. e. their concentration never reaches a high enough level for further nuclei to form. Slow production of growth units can be readily achieved by the controlled decomposition of a soluble Fe complex (see 4.2.5) or the dissolution of a solid precursor, such as 2-line ferrihydrite.

Methods used for the production of monodispersed Fe oxides include forced hydrolysis (i. e. hydrolysis at elevated temperatures) of acidic Fe^{III} solutions under well defined conditions (Section 4.2.1 and Section 10.2), the gel-sol method, seeding during forced hydrolysis and via solution transformation from either a solid or a soluble precursor. The seeding method was applied by Penners (1985) to forced hydrolysis of Fe^{III} solutions to produce monodispersed hematite crystals with sizes ranging from 0.15 to 0.56 μm. Most attention has been directed towards synthesis of hematite, although acicular goethite has been synthesized by hydrolysis of Fe^{III} solutions and uniform spheres of magnetite were produced by partial oxidation of $Fe(OH)_2$ at high pH (Sugimoto and Matijevic, 1980).

4.1.4 Production of Nanoparticles

Synthetic Fe oxides are usually in the micron or submicron size range. Ultrafine particles with dimensions in the nanometer range frequently require more specialized techniques than do the larger crystals. The two basic requirements for a monodisperse system must be met and in addition, coagulation and/or contact recrystallization of the primary particles must be prevented.

If precipitation is carried out a pH well away from the point of zero charge (PZC), coagulation can be prevented by repulsion of the electrical double layers of the particles. For maghemite nanoparticles, for example, precipitation is carried out at pH 2 (Tronc et al., 1995). Alternatively, a repulsive force, (involving electrostatic or steric hindrance) can be set up by adsorption of protective agents on the surface of the particles; for example, monodispersed iron oxide was produced by oxidation of Fe pentacarbonyl with H_2O_2 in ethanol. The alcohol helped stabilize the particles against coagulation (Joekes et al., 1981). Other workers have used gel networks or supports or vesicles to ensure separation of the particles. The supports can provide a framework or even a template for growth as well as preventing coagulation/agglomeration. Well dispersed hematite or maghemite nanoparticles have been prepared in silica xerogels (Morris et al., 1989; Chaneac et al., 1995; Ennas et al., 1998); such particles have been used to investigate magnetic phenomena. Well dispersed, nanosized magnetite has been precipitated in vesicles (Mann et al. 1979) and in elastic, polystyrene, polyacrylate copolymer gels (Breulmann et al., 1998). At present, these techniques rarely lead to a pure product. The Fe oxide is usually embedded in the support and cannot be separated from it. Such products are termed inorganic/organic composites. They are suitable for studies of magnetic and elastic properties. A related synthesis involves preparation of magnetic paper by in situ precipitation of maghemite and/or magnetite followed by introduction of the inorganic particles into the lumen of the cellulose fibres; excess of Fe oxide is washed off (Carrazaba-Garcia et al., 1997).

4.2 Main Routes of Synthesis

Almost all the iron oxides and hydroxides can be prepared by a number of synthesis pathways. The route followed frequently influences the properties of the product, particularly crystal morphology, degree of crystallinity, sample surface area and water content. The main synthesis methods are described in the following sections.

4.2.1 Hydrolysis of Acidic Solutions of Fe^{III} salts

At very low pH (< 1), Fe^{3+} exists as the purple, hexa-aquo ion, $[Fe(H_2O)_6]^{3+}$. Hydrolysis involves the stepwise elimination of protons from the six water molecules that surround the central Fe cation to form mono- and binuclear species. These species then interact further to produce species of higher nuclearity (Sylva, 1972; Johnston and Lewis, 1986; Bottero et al. 1994; Rose et al. 1997; Schwertmann et al. 1999). The latter finally precipitate as a more or less crystalline product, the nature of which depends on the rate and conditions of the reaction. Examples of hydrolysis and polymerization reactions are:

$$Fe^{3+} + H_2O \longrightarrow FeOH^{2+} + H^+$$

$$2Fe^{3+} + 2\,H_2O \longrightarrow Fe_2(OH)_2{}^{4+} + 2\,H^+$$

$$FeOH^{2+} + H_2O \longrightarrow FeOOH + 2\,H^+$$

$$2\,FeOH^{2+} + H_2O \longrightarrow Fe_2O_3 + 4\,H^+$$

Precipitation times may range from seconds to years. The two principal methods used to induce hydrolysis in the laboratory are, 1) heating the solution and 2) addition of base. As hydrolysis releases protons, the pH of the system falls during the early stages of hydrolysis. When hydrolysis is induced by raising the temperature, the pH may fall to a low enough value to inhibit further hydrolysis. This pH drop is important because incomplete hydrolysis reduces the yield of product.

Hematite, akaganeite, goethite and ferrihydrite may be obtained by hydrolysis of Fe^{III} solutions (Fig. 1-2). Which product forms is influenced by a variety of factors including rate of hydrolysis, solution pH, temperature, [Fe^{III}], and the nature of the anions present in the system. Unless the reaction conditions are carefully controlled, mixtures, rather than a single product, may result. Matijevic and his coworkers (Matijevic and Scheiner, 1978; Ozaki et al., 1984) have investigated this synthesis method intensively and shown the various ways in which the reaction product and also crystal morphology may be influenced by reaction conditions (Matijevic and Sapiesko, 2000).

The most important applications of the hydrolysis method are the production of hematite and akaganéite, but goethite can also be synthesized in this way. In the chloride system akaganeite may be a precursor of hematite (Hamada and Matijevic, 1981; Kandori et al., 1998). The main disadvantage of the method is a comparatively low yield of product due to the very low pH of the system.

4.2.2 Transformation of Ferrihydrite

2-line ferrihydrite is the initial product resulting from fast hydrolysis upon addition of alkali to a Fe^{III} salt solution. It is thermodynamically unstable and with time, transforms into goethite, hematite or a mixture of the two. Goethite and hematite form by different pathways hence conditions which hinder hematite promote goethite and vice versa. Hematite formation is promoted by factors that induce aggregation of ferrihydrite (pH near the PZC of ferrihydrite, i.e. pH around 8; Schwertmann and Murad, 1983) or dehydration (i.e. increased temperature). In contrast, goethite formation is favoured by raising or lowering the pH away from the PZC and by lower temperatures. Both processes appear to involve reconstructive via-solution reactions because the atomic arrangements of the two products are essentially different from that of ferrihydrite so that the two end products must be completely reconstructed. For goethite the reconstructive process takes place in the bulk solution, whereas formation of hematite must be preceded by aggregation of the ferrihydrite. Feitknecht and Michaelis (1962) suggested that hematite formation involves internal dehydration within the aggregate (single phase transformation) and is promoted by aggregation (Schwertmann and Fischer, 1966). The presence of an initial induction period during the transformation indicates that a nucleation stage is involved (Fischer and Schwertmann, 1975); nucleation requires the presence of at least some water within the aggregates. In fact, the water content in equilibrium with air (ca. 10%) was found to be sufficient for slow hematite formation at RT whereas ferrihydrite dehydrated to 5% water under vacuum was stable at 227 °C (Schwertmann et al., 1999). By using $\delta^{18}O$, it was recently shown, that the oxygen of the hematite formed from ferrihydrite came predominately from the water present during the transformation and not from the ferrihydrite precursor (Bao and Koch, 1999).

4.2.3 Oxidative Hydrolysis of Fe^{II} Salts

This is a very versatile method capable of producing goethite, lepidocrocite, magnetite, ferrihydrite and feroxyhyte. Which product forms depends on the reaction conditions. Careful control of pH, rate of oxidation, suspension concentration, temperature, and concentration of foreign species is needed to ensure that a pure product is obtained (Feitknecht, 1959; Schwertmann, 1959 b). In some cases an inert atmosphere of N_2 must be provided. Industrial pigments are usually produced by this method (Buxbaum, 1993).

The reaction can be carried out over the pH range 6–14. Between pH 6–7 goethite and lepidocrocite result: a pure product of either can be obtained by adjusting the rate of oxidation and the concentration of carbonate in the system (Schwertmann, 1959 b; Carlson and Schwertmann, 1990). At pH >8 magnetite is obtained and at pH 14, pure goethite is produced. With very rapid oxidation (e.g. by H_2O_2) feroxyhyte is obtained.

The oxidation/hydrolysis reaction releases protons, so that as the transformation proceeds and if no extra base is added, the pH drops to about 3 and the transformation rate falls to practically zero, thus leading to incomplete oxidation:

$$4\,Fe^{2+} + 6\,H_2O + O_2 \longrightarrow 4\,FeOOH + 8\,H^+$$

In order, therefore, to obtain a higher yield of reasonably well crystalline product, the pH must be held constant by continual addition of alkali to the system. This is most conveniently achieved by using an automatic burette and a pH-stat titration technique. The reaction is complete when no further addition of alkali is required. Alternatively, a buffer ($NaHCO_3$ or imidazole) may be used, but this will contaminate the end product. A feature of the oxidation/hydrolysis of Fe^{II} systems is that a crystalline product can be obtained within a few hours at room temperature.

4.2.4 Phase Transformations

Although synthesis of iron oxides from green rusts or from ferrihydrite involves phase transformations, the term is more commonly used to refer to conversion of one crystalline iron oxide or oxide hydroxide into another. Under suitable conditions, every iron oxide can be converted into at least one other member of the group (Mackay, 1961). Where a topotactic transformation* mechanism is involved, the product frequently retains the crystal morphology of the starting material (pseudomorphosis). This is the case, for example, for formation of maghemite by thermal dehydration of lepidocrocite (laths) or by oxidation of magnetite (cubes) and also for the formation of hematite by low temperature (200 °C) thermal dehydration of goethite.

4.2.5 The Gel-Sol Method

The gel-sol method involves aging a concentrated gel or iron polynuclear species at 100 °C for some days (Sugimoto and Sakata, 1992). The yield from this technique (in contrast to the forced hydrolysis method) is close on 100% and because a concentrated suspension is used, the yield can be considerable. This method has been applied mainly to haematite. It can also be used to produce small, uniform rods of akaganeite (Paterson and Tait, 1978 and Fig. 9.1 b).

* Topotaxy: Accord between the initial and product structures in three dimensions

4.2.6 Hydrothermal Precipitation

Hydrothermal precipitation involves a reaction in aqueous solutions at high temperatures (>100 °C) and pressures (1–3 Kbar). The reaction is carried out in an autoclave or a bomb. This method of crystal growth utilizes the fact that many substances that are insoluble in water at room temperature display appreciable solubility at high temperatures and pressures. In an autoclave goethite, for example, can be converted to hematite in alkaline media at high pressure at above 180 °C. At somewhat lower temperatures, multidomainic goethite recrystallizes to form single domain crystals (Schwertmann et al., 1985). Hydrothermal crystallization enables a product to be obtained in a far shorter time than by most other methods; the crystals are, in addition, much larger than usual – micron rather than submicron sized. Large hexagonal plates of hematite have been grown hydrothermally (Krahtovil and Matijevic, 1988) as have large crystals of magnetite (Viswanathiah and Krishnamurthy, 1980). Monodisperse micaceous iron oxide has been prepared by oxidation of iron under hydrothermal conditions (Uchida et al., 1993).

In the hydrometallurgical industry, Fe^{III} solutions are often produced as waste products and hydrothermal methods are being investigated as a possible route for their disposal by production of hematite of saleable quality (Riveros and Dutrizac, 1997).

4.2.7 Decomposition of Metal Chelates

Decomposition of Fe^{III} chelates is a suitable technique for production of fairly monodisperse products, often with unusual morphologies, because it provides a constant and controlled supply of reactant. The reaction is carried out at temperatures >100 °C and pHs >12. Oxidizing or reducing agents can be added to the system to direct the formation of a particular phase. Uniform, disk-like particles of hematite up to 15 µm across were produced by thermal decomposition of the Fe-triethanolamine (TEA) complex (H_2O_2 present) as were magnetite octahedra (hydrazine present) (Sapieszko and Matijevic, 1980). A disadvantage of this method is that the product may be contaminated by the ligand.

Cautionary Remark

In using the procedures described in the following sections it must be emphasized that small changes in reaction conditions i.e. pH, OH/Fe, Fe concentration, etc. may prevent the desired product from being obtained, hence absolute accuracy in following the preparative procedures is essential.

5 Goethite

5.1 Introduction

In laboratory studies goethite is one of the most widely used iron oxides. It serves as a model system for a great variety of investigations partly because its surface chemistry, surface structure and crystal morphology are well characterized and partly because it is the most widespread Fe oxide in natural environments. For this reason, some attention is given to production of goethites with different morphological characteristics. Synthesis of metal substituted goethites is also described; partial replacement of Fe in the structure by another metal ion can modify crystal properties such as morphology, size, crystallinity, solubility and color.

Goethite crystals are usually acicular and elongated along the crystallographic *a* direction (Figs. 5-1 and 5-2). (Note that the space group of goethite has been changed from Pbnm to Pnma and the former a, b, and c axes are now the c, a, and b axes (see also Tab. 1-3). Electron micrographs of thin sections show that the acicular crystals are bounded by {101} faces along the a-direction (Fig. 5-3), and terminate in {210} faces (Fig. 5-16). Well-formed acicular crystals, such as those in the AFM (Atomic Force Microscopy) picture of Fig. 5-5, also show (100) faces along the needle axis. Twinning is quite common (Fig. 5-4, top right). Goethite crystals often contain intergrowths or domains (Fig. 5-4, top left) which can accelerate dissolution of the crystals (Cornell et al. 1974).

Goethite may be synthesized from either Fe^{III} or Fe^{II} systems. In the Fe^{III} system goethite can form over a wide pH range. Methods of synthesis in both acidic and alkaline media are provided in the following section. In alkaline media, synthesis involves holding freshly prepared ferrihydrite (obtained by neutralizing a Fe^{III}salt solution with alkali) in KOH at pH >12 for several days. This method was first described by Böhm in 1925.

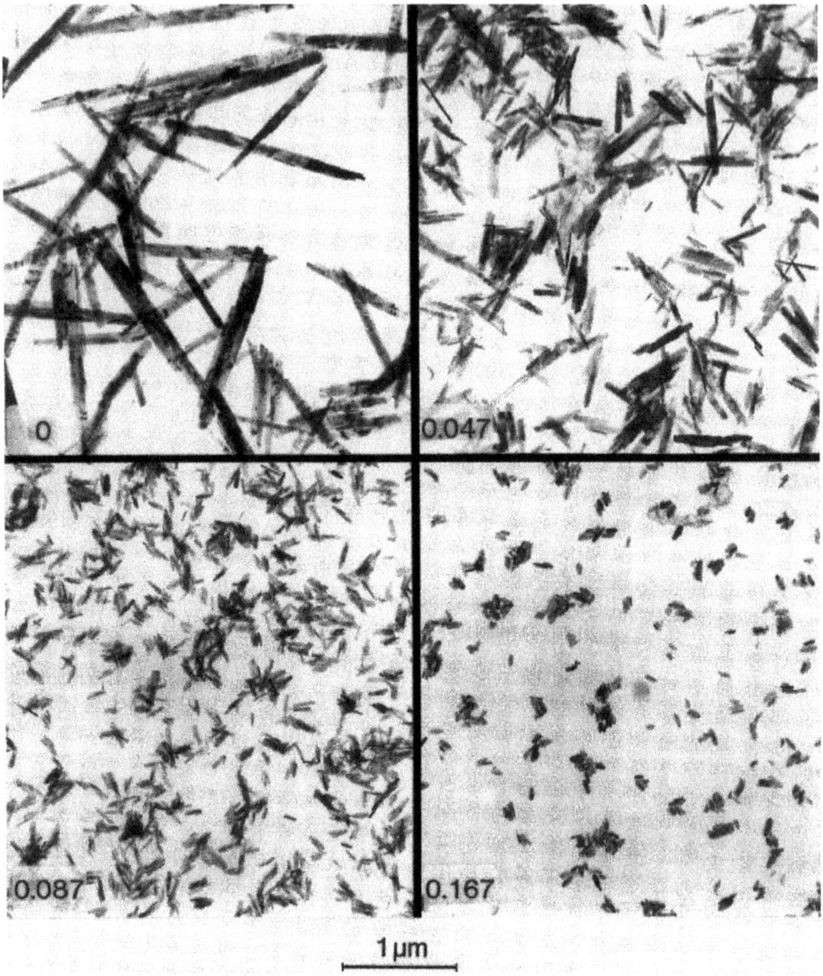

Fig. 5-1. Electron micrographs of Al-substituted acicular goethites illustrating the decrease in crystal size with increasing degree of Al-substitution (given as Al/(Al+Fe) mol/mol). The goethites were produced by aging 2-line Al-containing-ferrihydrites in 0.35–0.4 M KOH for 14 days at 70 °C (Cornell and Schwertmann; 1996; with permission)..

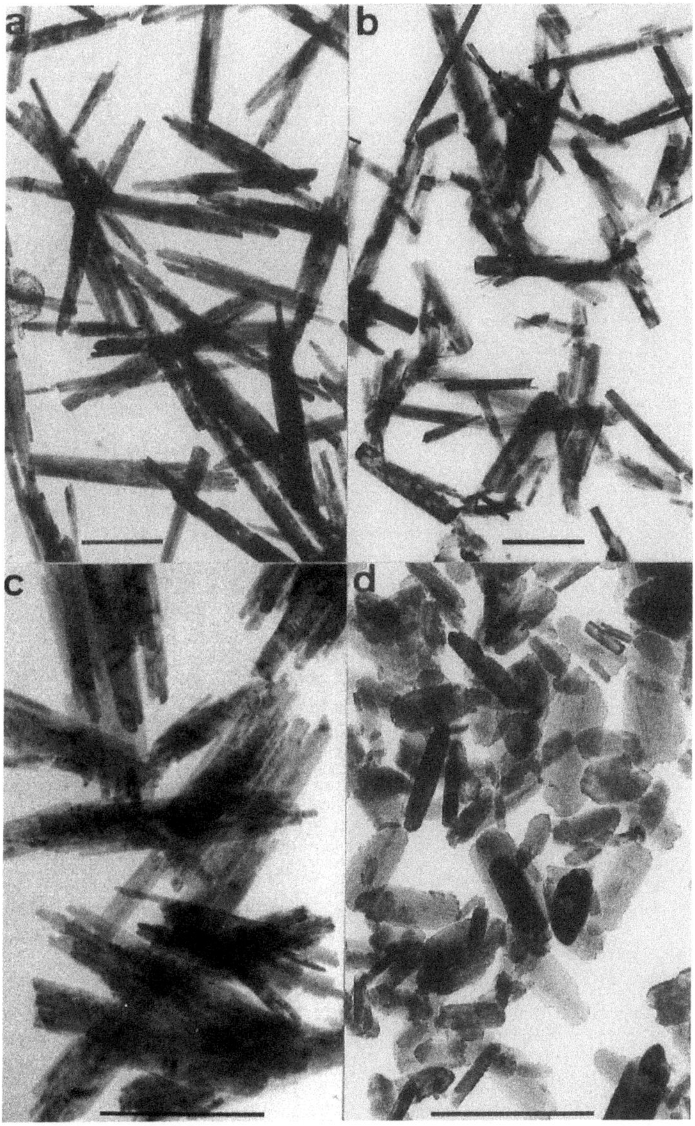

Fig. 5-2. Electron micrographs of pure and Al-substituted goethites grown in strongly alkaline conditions at 70 °C (a, b) or 25 °C (c, d). a: no substitution; b: 7.9 mol% substitution; c: no substitution; d: 11.6 mol% substitution. Bar = 100 nm; (see also Schulze and Schwertmann, 1984).

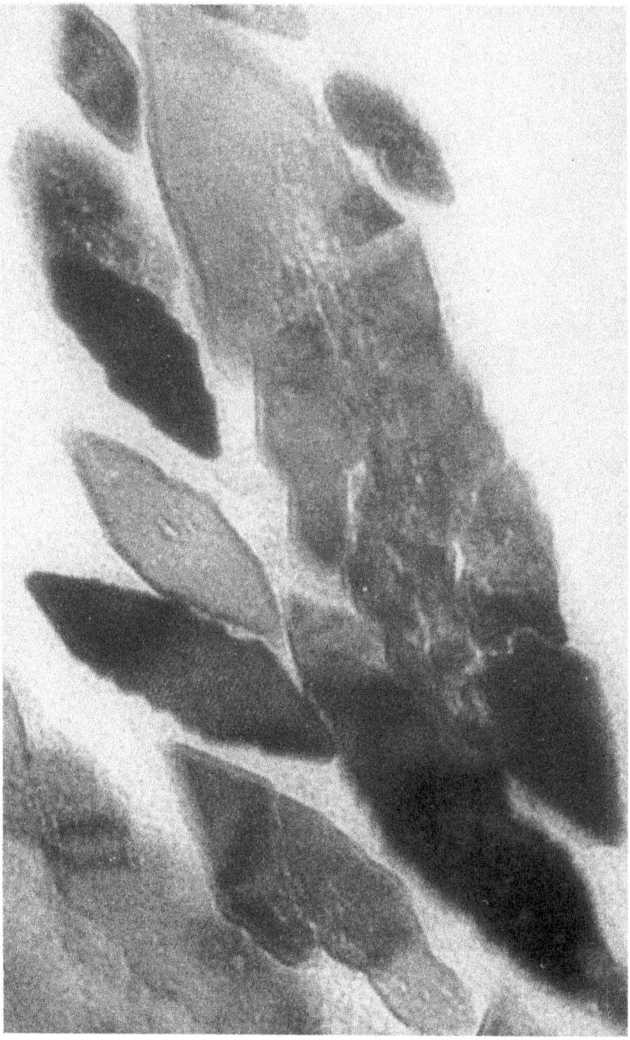

Fig. 5-3. High resolution electron micrograph of a thin section of acicular goethite crystals cut perpendicular to the crystallographic b-direction (needle axis). The lattice images (ca. 1 nm) seen in one crystal correspond to the a-dimension of the unit cell (0.9956 nm). The crystals are bounded by (101) faces. (Schwertmann, 1984; with permission). (Courtesy H. Vali).

Fig. 5-4. Crystal morphology of goethites. Cornell and Schwertmann, 1996; with permission.
Top left: Polydomainic crystals produced in 0.7 M KOH at 25 °C.
Top right: Polydomainic star-like twin produced in 0.3 M KOH at 70 °C (Courtesy P. Weidler).
Bottom left: Rafts consisting of short needles lying with their needle (*b*-) axis either parallel (a) or perpendicular (rhombic cross section) (b) to the plane of the paper. They were produced at RT from a partially neutralized $Fe(NO_3)_3$ solution at an initial pH of 1.7 (Courtesy S. Glasauer) (Method 5.2.2).
Bottom right: Subrounded crystals produced from 2-line ferrihydrite in the presence of cysteine (Method 5.2.3; for particle size distribution see Fig. 5-6).

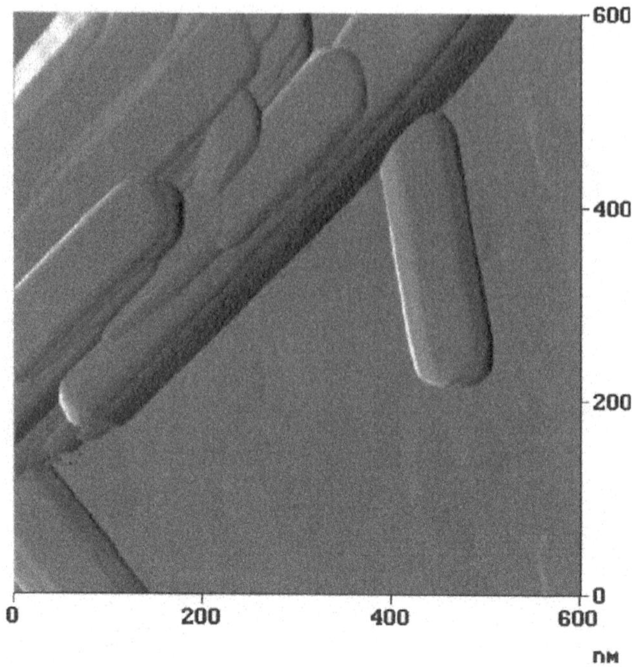

Fig. 5-5. Three-dimensional Atomic Force Micrograph of acicular goethites with (001) faces in the plane of the paper (courtesy P. Weidler).

Under alkaline conditions, ferrihydrite dissolves to release soluble Fe^{III} species ($Fe(OH)_4^-$) from which the less soluble goethite forms.

Under acidic conditions (pH 1–2) where the ferrihydrite precursor is highly charged and, thus, well dispersed, the soluble growth units from which goethite is most likely to be formed are mainly the monomer $Fe(OH)^{2+}$ and possibly the dimer $Fe_2(OH)_4^{2+}$ (Sylva, 1972; Johnston and Lewis, 1986; Cornell et al. 1989; Schwertmann et al. 1999). Precipitation is much slower (weeks to months at RT; see 5.2.2) than under alkaline conditions.

Both goethite and hematite can form from ferrihydrite. Under alkaline conditions, hematite formation is less likely the higher the pH and the lower the temperature (Schwertmann and Murad, 1983; Cornell and Giovanoli, 1985). At $<70\,^{\circ}C$, a pH >12 is usually sufficient to avoid hematite (Lewis and Schwertmann, 1980).

The widely used method of Atkinson et al. (1968) is essentially a combination of the alkaline (Böhm) and the acidic (Mørup) methods. It involves pre-aging a $Fe(NO_3)_3$ solution with an OH/Fe ratio of 2 (pH ca. 1.5) at room temperature, before making the system alkaline (pH ca. 12) and holding it at 60 °C. Although it has not been investigated in detail, it is probable that the 50 day pre-aging step leads to the formation of small, acicular goethite crystals which increase in size during the subsequent ageing at high pH and temperature. Thus, this first step is not essential, because very similar crystals are obtained whether it is included or not.

Synthesis from Fe^{II} systems involves oxidative hydrolysis of Fe^{II} solutions. The initial precipitate may be a so-called green rust (Bernal et al., 1959; Taylor, 1980). As various Fe oxides (lepidocrocite, goethite, feroxyhyte and magnetite) may be produced by this method, careful control of factors such as the rate of oxidation, pH and the nature of the anion present is necessary to ensure formation of pure goethite.

Production of goethites in the structure of which Fe has been partly replaced by trivalent Al, Mn, Cr or V (= M), follows, in principle, the same method as that used for unsubstituted goethite, except that the Fe solution is replaced by mixed Fe-M solutions. By varying the M/(M + Fe) ratio different levels of substitution may be obtained. Crystallization is often retarded by the presence of foreign cations, e.g. Co, Ni, Cu and Zn (Giovanoli and Cornell, 1992) and Ni and Pb (Ford et al. 1999), so transformation times are usually longer than for pure goethite. The retarding effects of Ni and Pb (Ni > Pb) were the inverse of the surface stability constants of the two metals (Pb > Ni).

5.2 Pure Goethite from Fe^{III} Systems

5.2.1 Preparation from an Alkaline System (acc. to Böhm, 1925)

Method

Solutions: A) 1 M $Fe(NO_3)_3$ solution freshly prepared by dissolving unhydrolysed $Fe(NO_3)_3 \cdot 9\,H_2O$ in twice distilled water

B) 5 M KOH

Pour 100 mL of solution A into a 2 L polyethylene flask, and add rapidly and with stirring, 180 mL solution B. Red-brown 2-line ferrihydrite precipitates at once. Immediately dilute the suspension to 2 L with twice distilled water and hold in a closed polyethylene flask at 70 °C for 60 hr. The preparation must be carried out in polyethylene vessels because in strongly alkaline media, some Si dissolves from glass vessels (see below). During the heating stage, the voluminous, red-brown suspension of ferrihydrite is converted to a compact, yellow brown precipitate of goethite. Remove the reaction vessel from the oven, centrifuge, wash and dry the precipitate.

Washing the product is essential to remove OH^- and NO_3^-: The presence of the latter ion in solution is tested qualitatively with diphenylamine. In the product it can be recognized by a sharp IR-peak at 1400 cm^{-1}.

Product Description

This method produces ca. 9 g goethite with a surface area of ca. 20 m^2/g. The crystals are acicular and consist of several domains along the needle (*a*-)axis. An electron micrograph is shown in Fig. 5-2 a, an X-ray diffractogram in Fig. 5-6 and an IR spectrum in Fig. 5-7. The crystals are bounded mainly by {101} faces (Fig. 5-3).

5.2.2 Preparation from an Acid System (acc. to Mørup et al., 1983)

Method

Dissolve 283 g of $Fe(NO_3)_3 \cdot 9\,H_2O$ in 350 ml 2M HNO_3, dilute with distilled water to 1.4 L and add 1.4 L of 1 M NaOH (OH/Fe = 2.0) with vigorous stirring. The pH is 1.7–1.8. A stable brown sol of ferrihydrite forms from which yellow goethite starts to precipitate after 50 d (see below). Centrifuge, wash free from electrolyte and freeze dry.

Product Description

The goethite consists of fairly monodisperse acicular crystals about 20–25 nm long, 10–15 nm wide and 4–10 nm thick (Fig. 5-4; bottom, left). The diamond-shaped cross section indicates that the crystals are again essentially bounded by {101} faces. The surface area is ca. 130 m^2/g and the magnetic hyperfine field at RT is 38–39 T (for further properties

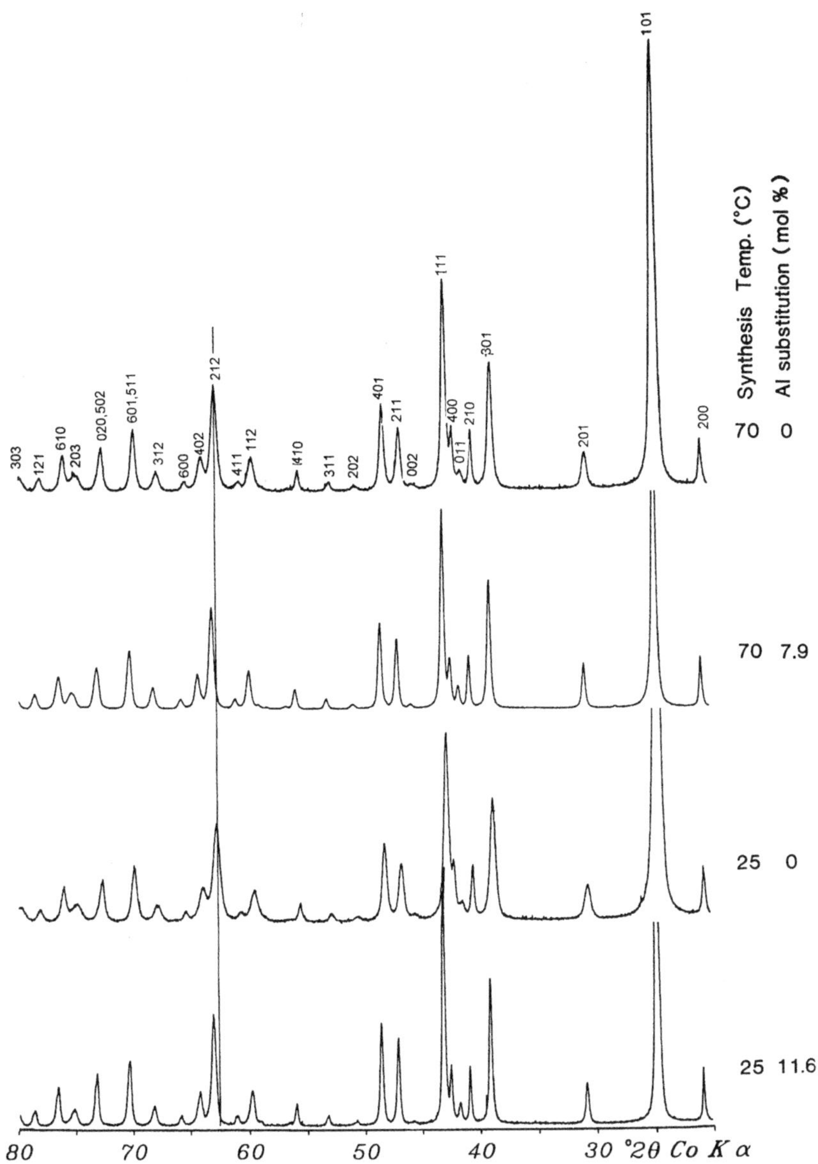

Fig. 5-6. X-ray diffractograms of pure and Al-substituted goethites prepared in 0.3 M KOH at 70 or 25 °C. The vertical line at the 212 peak illustrates the peak shift to higher angles as Al-for-Fe substitution increases.

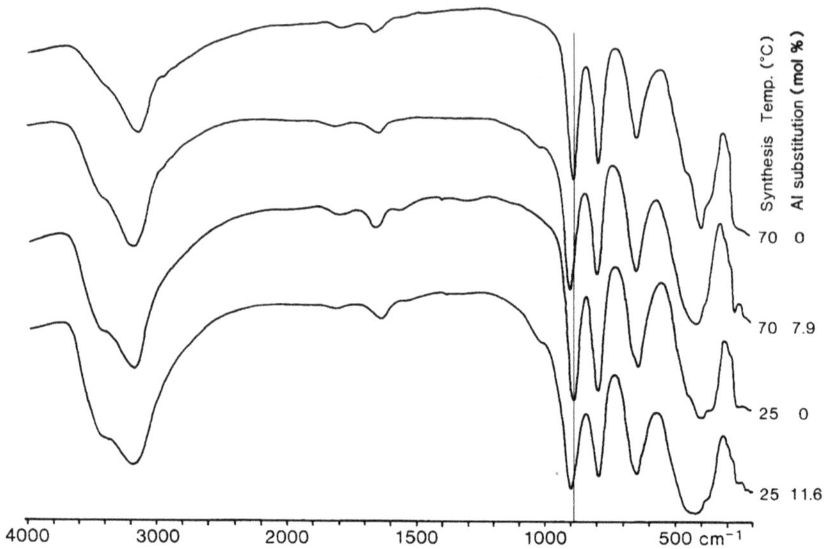

Fig. 5-7. Infra red spectra of pure and Al-substituted goethites prepared at 25 and 70 °C. The vertical line illustrates the shift in the position of the OH bending mode due to Al-for Fe substitution.

see Atkinson et al. 1968, 1977; Mørup et al., 1983; Glasauer et al. 1999). The significantly smaller crystal size and the correspondingly higher surface area as compared to those of the goethites produced under alkaline conditions (see 5.2.1) can be attributed to the higher nucleation rate under acidic conditions. As precipitation at room temperature is slow, a yield of ca. 25% is obtained after 50 days (ca. 10 g) and 40% after 660 days. First order precipitation kinetics are followed.

5.2.3 Preparation from a Cysteine/2-line Ferrihydrite System

Method

Dissolve 10 g $Fe(NO_3)_3 \cdot 9\ H_2O$ in 650 ml twice distilled water in a 1 L polyethylene flask, add rapidly and with stirring 70 mL M KOH, followed by 50 mL of 0.045 M $NaHCO_3$ to buffer the ferrihydrite suspension at pH 8.5. Then add with stirring a freshly prepared solution of 3.5 g

L-cysteine in 30 ml of water to give a cysteine/Fe ratio of 1. The red-brown ferrihydrite rapidly darkens. Close the flask and hold at 70 °C for 60 hr after which a compact, black precipitate will have formed. Shake the flask vigorously to disperse the precipitate, then aerate the suspension with air at a rate of 40 L/hr through a glass tube for 6 hr. During oxidation the suspension cools to room temperature and the color changes from black to yellow-brown. Although the yellow color develops after one hr, the goethite at this stage is poorly crystalline. Wash and dry the end product in the usual manner. During the ageing of the ferrihydrite-cysteine suspension, a head space of ca. 6 cm must be maintained. However, oxidation of the system before the black phase has formed must be avoided.

The cysteine/Fe ratio must not fall below 1 or reasonably uniform crystals will not form. As cysteine is readily oxidized even in air, a fresh solution must be prepared just before starting the preparation. Solid cysteine also tends to oxidize gradually once the container has been opened, with the result that with time, the calculated cysteine: Fe ratio may be slightly <1. Using NH_4Cl/NH_3 buffer instead of $NaHCO_3$ induces formation of acicular goethite crystals up to 150 nm long (similar to those in 5.3.1).

Product Description

The method produces ca 2.25 g fairly uniform, subrounded crystals with an average length of 30–40 nm (Fig. 5-4, bottom right), and a surface area of ca. 30 m^2/g. The frequency distribution of crystal lengths is shown in Fig. 5-8. Chemical microanalysis showed <2% C, S, or N in the product.

Comments

The black compound which is oxidized to goethite has large platy crystals (surface area ca. 10 m^2/g), shows two strong basal reflections at 1.04 and 0.504 nm, and has an Fe^{II}/Fe^{III} ratio of ca. 0.8–1. Thus, it appears to be a hitherto unknown layer compound akin to the well-known green rust phases (see chap. 13).

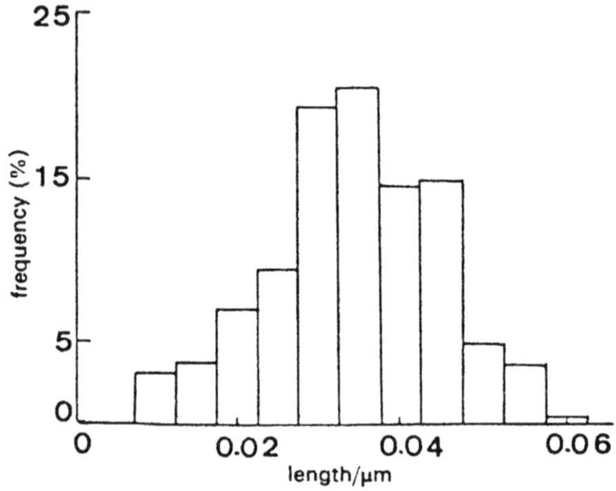

Fig. 5-8. Frequency distribution of lengths of goethite crystals produced by the cysteine method (method 5.2.3) (Cornell et al. 1991. Reproduced by permission of the Royal Society of Chemistry.).

5.3 Pure Goethite from an Fe^{II} system

Method

Dissolve 9.9 g unoxidized crystals of $FeCl_2 \cdot 4H_2O$ or 13.9 g $FeSO_4 \cdot 7H_2O$ in 1 L of distilled water through which N_2 has bubbled for 30 min beforehand in order to remove dissolved oxygen. The solution should be held in a wide-mouthed, 2 L bottle. Add 110 ml of 1 M $NaHCO_3$ solution and replace the N_2 purge gas by air with a flow rate of 30–40 ml/min. Stir continuously. Oxidation is complete within 48 hrs during which time the color of the suspension changes from green-blue to ochreous. The pH during oxidation is maintained at about 7 by the $NaHCO_3$ buffer.

Product description

The method produces ca. 5–6 g goethite of low crystallinity and a surface area of about 80 m^2/g. The crystals resemble goethites from various natural environments. An X-ray diffractogram is shown in Fig. 5-9, an EM photo in Fig. 5-10 and a Mössbauer spectrum in Fig. 5-11. An X-ray

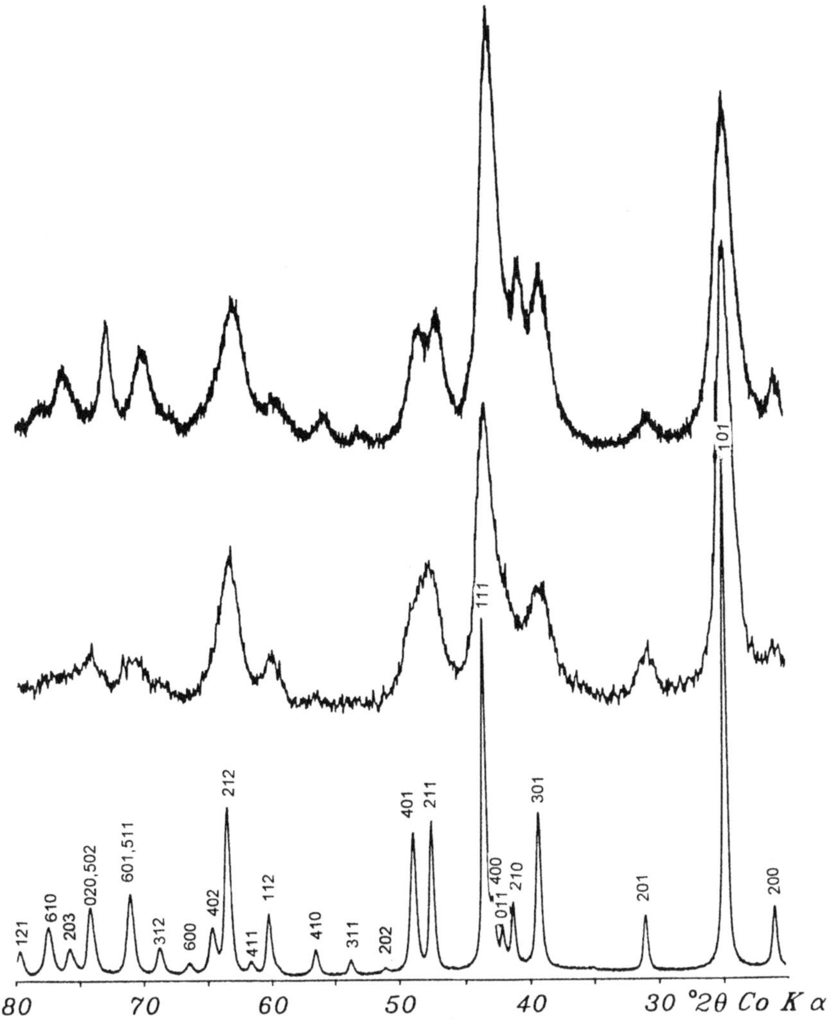

Fig. 5-9. X-ray diffractograms of goethites with and without Al substitution produced from Fe^{II}systems. Top: Poorly crystalline pure goethite synthesized at room temperature by oxidation of $FeCl_2$ solution in the presence of carbonate at pH 6–7. Middle: As before but with 31 mol% Al substitution. Bottom: Well crystalline goethite with 30 mol% Al substitution synthesized by oxidation of a $FeCl_2$ solution at room temperature and pH 11.6.

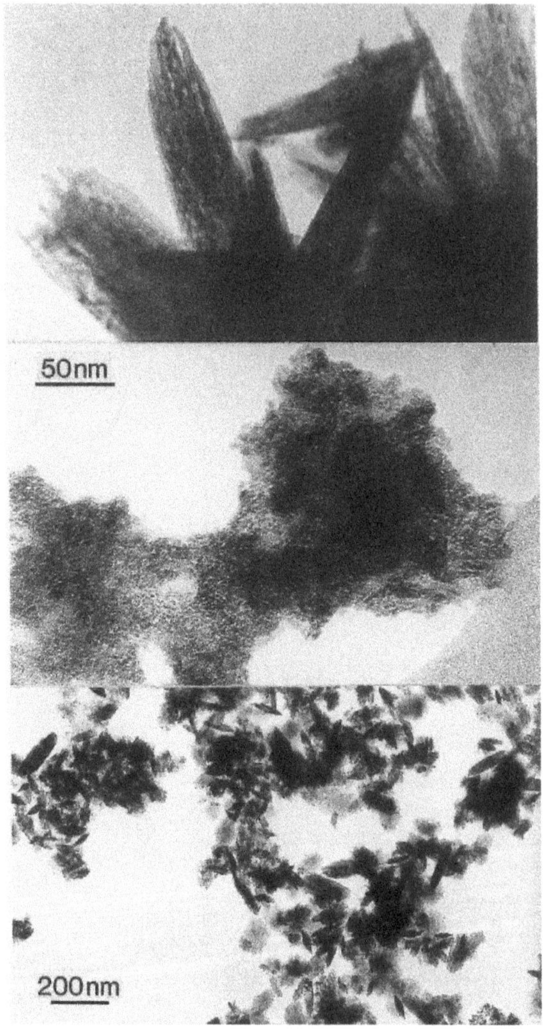

Fig. 5-10. Electron micrographs of goethites with and without Al substitution produced from Fe^{II} systems. For explanation see Fig. 5-9.

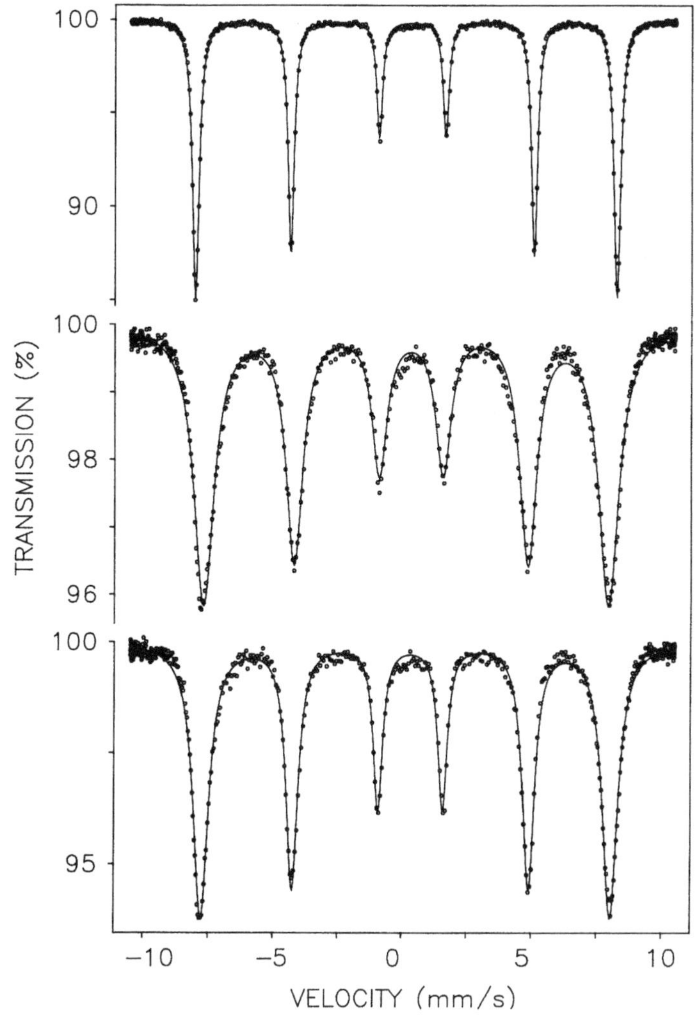

Fig. 5-11. Mössbauer spectra obtained at 4.2 K of goethites with and without Al substitution produced from Fe^{II}systems. For explanation see Fig. 5-9.

photoelectron spectrum characterizing the binding energy of the surface Fe and O atoms at the (100) face of this goethite was obtained by Rakovan et al. (1999). The goethite contains some tenth of a per cent of carbon in the form of adsorbed carbonate as shown by two broad IR bands at around 1300 and 1500 cm^{-1} (Fig. 5-12). The carbonate appears to be strongly adsorbed and is therefore difficult to remove, e. g. by washing.

The presence of carbonate suppresses the formation of lepidocrocite (Schwertmann, 1959 b; Goodman and Lewis, 1981; Carlson and Schwertmann, 1990). For this purpose a mole ratio $[HCO_3]/[Fe]$ of 1.5–2 is required. The presence of lepidocrocite (arising from too low a level of carbonate) may be indicated by an orange tinge.

If the rate of oxidation is substantially higher than indicated above, some lepidocrocite may form. Without thorough agitation of the whole suspension some (black) magnetite may form in regions of locally higher pH. The crystallinity of goethite from Fe^{II} sulfate is somewhat higher than that from Fe^{II} chloride.

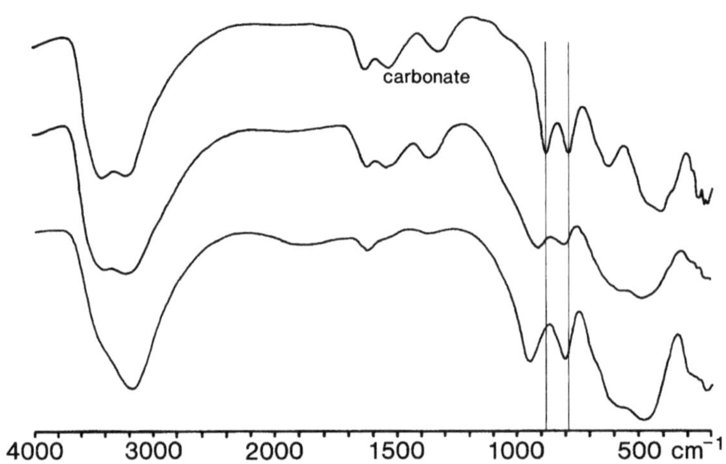

Fig 5-12. Infra red spectra of goethites with and without Al substitution produced from Fe^{II}systems. For explanation see Fig. 5-9. The upper two goethites contain some carbonate. The vertical line shows the shift in the positions of the OH bending mode at 800 and 900 cm^{-1} due to Al-for-Fe substitution.

5.4 General Comments

Although pure goethite can be easily obtained over a wide temperature range (4–70 °C) under strongly alkaline conditions, both *crystal size* (ca. 0.01–1 μm) and *shape* vary widely and this offers the possibility of producing a range of different goethites for experimental purposes. The common shape of goethite crystals is the needle (acicular). The length/width ratio (aspect ratio) of the needles varies widely: the crystals can range from long, thin needles to short, broad blocks. This is the result of variations (due to changes in precipitation conditions) in the ratio of the crystal growth rate along the needle axis to the rate in the other directions, i. e. along the *a* direction vs. along the *b* and *c* directions. Some experimental results on this subject are summarized in the following section.

In 0.3 M [OH^-] (as in the above recipe) the crystal size of goethite decreased as the *temperature* fell from 70 to 4 °C and the surface area increased by an order of magnitude (13→153 m^2/g) (Schwertmann et al. 1985). At the same time the rate of formation slowed down. At any particular temperature, the crystals become smaller as the *OH^- concentration* increases. This is because with increasing [OH^-], the supersaturation with respect to goethite and, in turn, the rate of nucleation, increase. For example, in 0.3 and 2M [OH^-], surface areas of 16 and 88 m^2/g, respectively, were obtained (Schulze and Schwertmann, 1984). Extremely thin and long (>1 μm) needles, either single or multidomainic, are obtained at pH >13, whereas thin, but short needles (see above) formed at low pH (Fig. 5-4) (Schwertmann, 1965; Atkinson et al. 1968; Mørup et al. 1983; Glasauer et al. 1999). Blocky twinned crystals occurred at pH 11–12 (Atkinson et al. 1968). The goethites produced by oxidation of Fe^{II} at RT, which is most likely to be the dominant route of formation in surface environments, are also acicular and polydomainic but less well developed (Fig. 5-10). Highly multi-domainic crystals (Fig. 5-4) formed in the pH range 13–13.5 (Cornell and Giovanoli, 1986). As the pH fell the domainic character decreased and goethites grown at pH 11.5 were predominantly single domain. Domainic character also increased as the temperature of synthesis fell from 70 °C to 4 °C. The domains can be eliminated from the low temperature goethites by healing the crystals in water in a pressure vessel at 180 °C (Fig. 5-13) (Schwertmann et al., 1985).

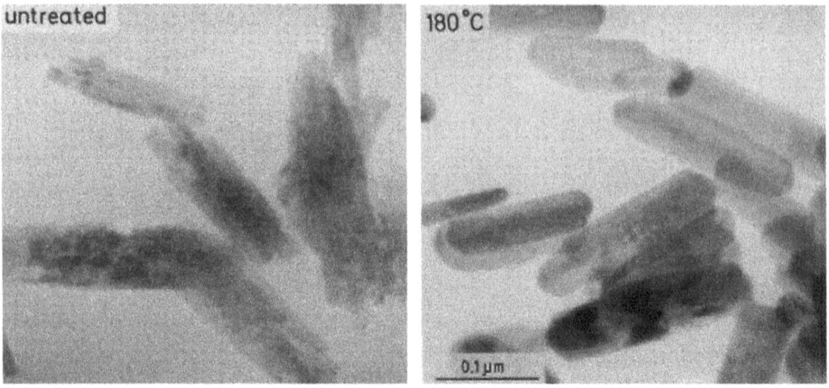

Fig. 5-13. Polydomainic goethite crystals formed in 0.7 M KOH at 4 °C (left) transformed into monodomainic, smooth crystals after hydrothermal treatment at 180 °C (right) (Schwertmann et al., 1985; with permission).

Twinned crystals of different kinds have been obtained under various conditions (Cornell and Giovanoli, 1985). Twin pieces and dendritic twins formed from concentrated suspensions over the pH range 11.5–12 at 70 °C and at higher pH were induced by high ionic strength or foreign, adsorbed ions (e.g. Mn). Star-shaped twins (Fig. 5-4 top right) form in very concentrated suspensions between pH 12–13. Twinned crystals are always associated with acicular crystals.

Goethite is usually prepared in unstirred systems. Continuous *stirring* during growth led to larger crystals than those obtained in unstirred systems (e.g. SA: $38 \rightarrow 8\ m^2/g$) and to improved crystallinity as seen from a significant decrease in the unit cell edge length c from 0.4618 to 0.4611 nm (Schwertmann and Stanjek, 1998). This deviation of c from the ideal value (0.4608 nm) was attributed to some extra structural OH due to structural defects and/or Fe-vacant sites in poorly crystalline products. An even more perfect goethite (c-value of 0.4608 nm) has been obtained by *hydrothermal* synthesis (155 °C; 2 M [OH^-]) (Thiel, 1963).

Silicate retards the conversion of ferrihydrite to goethite and small amounts of Si are retained by the goethite (Atkinson et al., 1968; Cornell et al., 1987). With high levels of silicate (Si/Fe = 0.1), larger crystals with pseudohexagonal or bipyramidal shapes result (Cornell and Giovanoli, 1987a; Glasauer et al. 1999). The larger size may be due to hindered nucleation (fewer crystals) and the modified shape due to preferential adsorption of silicate ions on the terminal faces of the crystals thus retarding their growth and promoting their development.

Hematite is the phase most likely to contaminate goethite preparations from Fe^{III} systems. Mixtures of goethite and hematite may form outside the range 0.1–4 M KOH (Fig. 5-14) and also at temperatures above 80 °C, even at pH 12. The presence of hematite can be recognized by the orange to red color of the product which depends on the proportion of hematite present. Epitaxial twins consisting of hematite centres with up to six outgrowths of goethite (Fig. 5-15) are obtained when a proportion of hematite forms as well as goethite; usually in the pH range 10.5–11.5.

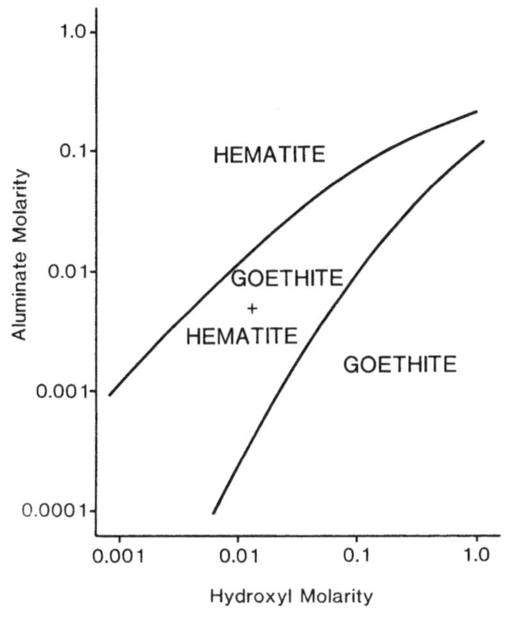

Fig. 5-14. Relation between extent of hematite and goethite formation and the hydroxyl and aluminate molarity at 70 °C (Lewis and Schwertmann, 1979; with permission)

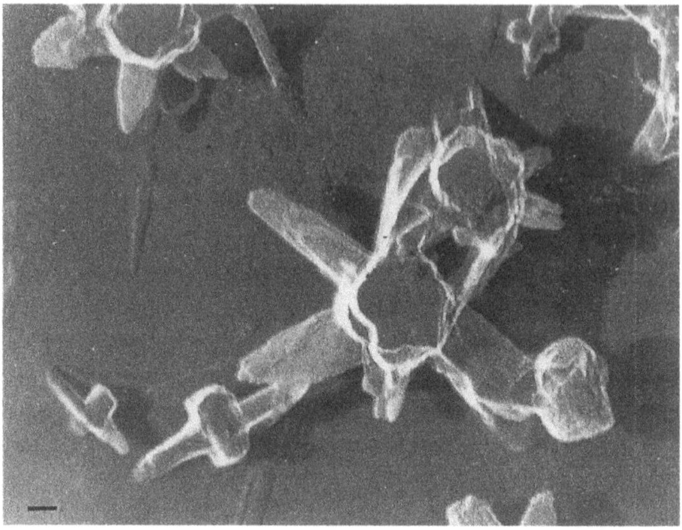

Fig. 5-15. Electron micrograph replica of acicular goethite outgrowths on a hematite core. The crystals grew from ferrihydrite at pH 11.7 and 70 °C in the presence of 10^{-4} M maltose in 24 h. Bar = 100 nm. (see Cornell, 1985).

5.5 Metal (M)-Substituted Goethites $Fe_{1-x}M_xOOH$

5.5.1 Al-Substituted Goethite $Fe_{1-x}Al_xOOH$

5.5.1.1 Preparation from an Alkaline Fe^{III} system

Solutions: A. 1 L of 1 M $Fe(NO_3)_3$ solution.
B. 500 mL 0.5 M $Al(NO_3)_3$ solution.
C. 2 L of 5 M KOH
D. Aluminate solution (0.3125 M) prepared by addition of solution B to 300 mL of solution C with constant stirring. The final solution should be clear. Solutions C and D must be held in polyethylene flasks.

To produce a series of seven goethites (approximately 9 g each) with between zero and about 10–12 mol % Al substitution add 0, 10, 20, 30,

50, 80 and 120 mL of soln. D to 2 L polypropylene bottles, followed by 180, 179, 178, 176, 174, 170 and 165 mL respectively of soln. C to the bottles. Then *quickly* add 100 mL of freshly prepared soln. A to each bottle and make up to 2 L with distilled water. The corresponding Al/Fe+Al mole ratios in solution are 0; 0.03; 0.0588; 0.0857; 0.135; 0.20; and 0.2728 mol/mol. Mix thoroughly and place in an oven at 70 °C for about 14 days. Shake once a day. After crystallization is complete (as indicated by a compact, yellowish-brown precipitate or a low Fe_o/Fe_t ratio) centrifuge, wash twice with 400 mL M KOH to remove extra Al, adjust to pH 7.5 with M HCl, wash with distilled water and dry.

Product description

The surface area of the goethites is between 20 and 30 $m^2\ g^{-1}$ and increases slightly with increasing Al substitution. Because of the high solubility of Al hydroxides at high $[OH^-]$, the maximum Al substitution of up to ca.16 mol% is below the maximum possible level of 33 mol% even if the initial Al/(Al+Fe) in solution is much higher; various $Al(OH)_3$ phases (gibbsite, bayerite) will form as well. The crystals are acicular. With increasing Al substitution the needles become shorter and less multidomainic (Schulze and Schwertmann, 1987) (Fig. 5-1) and the surface area increases. X-ray diffractograms (Fig. 5-6) show a shift of all peaks towards lower d-values which can be used for quantifying the extent of unit cell diminution and, hence of Al-for-Fe substitution even in mixtures of goethite with hematite. A corresponding shift seen in the IR-absorption bands (Fig. 5-7) is again related to substitution. Due to the diamagnetic character of Al, the magnetic hyperfine field decreases with increasing Al-substitution (Schwertmann and Murad 1983).

Comments

The extent of Al substitution is primarily a function of the Al activity in solution but other factors also play a role. For example, at 70 °C and an initial Al/(Al+Fe) of 0.20 mol/mol in solution, substitution decreased from 15 to 4 mol% as the $[OH^-]$ increased from 0.1 to 1 M. Temperatures of between 25 and 90 °C had essentially no effect if the initial Al/(Al+Fe) was ≥ 0.5 but at ≤ 0.33, substitution decreased as the temperature rose from 25 °C to 70 °C (Lewis and Schwertmann, 1979; Cornell and Schwertmann, 1996). Al in the system favored hematite over goethite formation. This effect increased with increasing temperature and, at a given temperature (70 °C), and in the $[OH^-]$ range of 10^{-3}–1 M, it decreased with increasing $[OH^-]$ (Fig. 5-14). At RT and in the pH range 4–7, 2.5 mol% Al in the system was sufficient to suppress goethite formation completely in favor of hematite even in liquid water (water activity = 1) (Schwertmann et al. 2000). Thus, hematite formation can be avoided either by lowering the temperature (causing crystallization to proceed more slowly) or by increasing $[OH^-]$ (reducing Al substitution). Stirring during synthesis increased Al substitution at any given initial Al/(Al+Fe) and favored hematite formation slightly (Schwertmann and Stanjek, 1998).

5.5.1.2 Preparation from an Fe^{II}-Sytem

Because no more than ca. 16 mol% Al substitution can be obtained in an alkaline Fe^{III} system, a recipe with which ca. 30 mol% substitution can be obtained by using Fe^{II} as an Fe source is described below.

Method
Add 50 mL of freshly prepared M $FeCl_2$ solution and 25 mL of M $AlCl_3$ solution to a 2 L polypropylene bottle, dilute to approximately 1 L and adjust the pH to about 11.7 with M KOH. Mix thoroughly. Slowly oxidize the suspension by opening the bottle daily and swirling the contents. Oxidation is completed after approximately 2–3 months when a dense yellow product has formed. Centrifuge and wash twice with 0.01 M KOH to remove the excess of Al, then wash with water and dry.

Product Description
This method produces 5 g of reasonably well crystalline goethite with Al substitution of ca. 30 mol%. The crystals are acicular/somatoidal in shape in contrast to the acicular form of those produced under strongly alkaline conditions from Fe^{III} systems. The surface area is around 35 m^2/g. An X-ray diffractogram is shown in Fig. 5-9 bottom, an electron micrograph in Fig. 5-10 bottom, an IR spectrum in Fig. 5-12 bottom and a Mössbauer spectrum in Fig. 5-11 bottom.

Comments

A lower degree of Al substitution can be obtained by this method with a lower initial Al/(Al+Fe) ratio. However, under these conditions, Al substituted magnetite is formed in addition to goethite (Schwertmann and Murad, 1990). The goethite may be purified by extracting the magnetite with M H_2SO_4 at 80 °C.

High-Al goethites may also be produced from mixed $FeCl_2$-$AlCl_3$ solutions by the same method as that described for pure goethite. Such Al goethites are, however, very poorly crystalline and have surface areas which vary from 100 to 300 m^2/g as Al substitution rises (Goodman and Lewis, 1981; Murad and Schwertmann, 1983). An X-ray diffractogram, electron micrograph, Mössbauer and IR spectrum are shown in Figs. 5-9 (middle), 5-10 (middle), 5-11 (middle) and Fig. 5-12 (middle), respectively.

5.5.2 Cr-Substituted Goethite $Fe_{1-x}Cr_xOOH$

Method

Mix 600 ml of 0.5 M $Cr(NO_3)_3$ solution with 360 ml of 5 M KOH in a 2 L plastic bottle. Add 90 ml of 1 M $Fe(NO_3)_3$ solution to 0; 9; 18; 27; 45; 72; 108; 135; 180; and 270 mL of the alkaline Cr solution and bring to 0.3 M KOH with 162; 161; 160; 159; 156; 153; 148; 145; 140; and 128 mL of 5 M KOH in a final volume of 1.8 L. Store the suspension in an oven at 70 °C for 111 days and shake occasionally. Wash the product twice with 0.1 M KOH and then with distilled water to remove the electrolytes and dry. Remove X-ray amorphous Cr-hydroxides and adsorbed Cr by a 2 hr treatment with 2 M H_2SO_4 at 80 °C (Schwertmann et al., 1989).

Product Description

The Fe_o/Fe_t ratio of the product increases from 0.002 to 0.057 as the Cr/Cr+Fe in the system increases indicating nearly complete transformation to goethite in all cases. The Cr-for-Fe substitution ranges from between 0 and 12 mol%, i.e. not all the Cr added to the system is incorporated into goethite. The goethites becomes considerably darker as substitution increases (see Plate VI). The size of the acicular crystals decreases with increasing substitution and resistance to dissolution in 6 M HCl at RT increases. The pure Cr hydroxide (no Fe addition) obtained under these conditions is amorphous.

5.5.3 Mn-Substituted Goethite $Fe_{1-x}Mn_xOOH$

Method

Add 175 mL of 2 M NaOH to 50 mL of mixed solutions of $Fe(NO_3) \cdot 9H_2O$ and $Mn(NO_3)_2 \cdot 4H_2O$ with a total (Fe + Mn) concentration of 0.53 M and Mn/(Mn+Fe) ratios of up to 0.10 mol/mol. Centrifuge and store the dark precipitates in polyethylene bottles in 250 mL 0.3 M NaOH at 60 °C for 15–20 days. Centrifuge, wash and dry (Stiers and Schwertmann, 1985; Cornell and Giovanoli, 1987).

Product description

The product is essentially fully crystalline ($Fe_o/Fe_t < 0.01$). Goethite is the only phase at an initial Mn/(Mn+Fe) of 0–0.10 mol/mol and contains up to 10 mol% Mn. Mn-substituted goethites are darker than unsubstituted goethites and have an olive tinge (see Plate VI). The Munsell color notation goes from 10YR 7/8 to 2.5Y 3/2 as the substitution increases from 0 to 12.5 mol% (see Fig. 3-7). Spot analytical electron microscopy (AEM) showed that the Mn content in the core of single crystals was higher than in the tip (Gasser et al. 1999). Whereas all three unit cell edge lengths of Al- and Cr-substituted goethites decrease with increasing substitution, for Mn-substituted goethite only *b* and *c* decrease while *a* increases. This result supports the assumption made on structural grounds that although added as Mn^{2+}, the Mn was oxidized and incorporated into the structure as Mn^{III}. Further evidence for trivalency of Mn in the structure comes from X-ray absorption spectroscopy (EXAFS) (Scheinost et al. et al. 1999). The presence of Mn in the system promotes dendritic twinning of goethite particularly at Mn/(Mn + Fe) ratios above 0.1 mol/mol.

Comments

At high enough Mn/(Mn+Fe) ratios, Mn phases may form in these systems. Cornell and Giovanoli (1987) reported that under alkaline conditions, increasing amounts of the spinel, jacobsite, ($MnFe_2O_4$), formed if the initial Mn/(Mn+Fe) >0.1. Formation of jacobsite was also promoted by increasing the suspension concentration. Since jacobsite is black, large amounts of this compound give the product a black appearance. Jacobsite and hausmannite were found at an initial Mn/(Mn+Fe) of >0.4 at pH 4, 6, and 8 at 50 °C after 40 days (Ebinger and Schulze, 1990) and groutite, the MnOOH form isostructural with goethite, formed at pH 4 and 6 after storage of a mixed precipitate with an initial Mn/(Mn+Fe) of 0.50 at 55 °C for 62 days (Ebinger and Schulze, 1989). At 70 °C, hausmannite and goethite formed at an initial Mn/(Mn+Fe) >0.2 at pH 8 after several weeks (Cornell and Giovanoli, 1990). A linear decrease in the magnetic hyperfine field at RT from 38.3 to 34.8 T was reported for Mn-substituted goethite when the Mn/(Mn+Fe) ratio increased from 0 to 0.09 mol/mol (Vempati et al. 1995).

5.5.4 V-Substituted Goethite $Fe_{1-x}V_xOOH$

Method

Add 0; 0.903; 2.306; 3.539; or 6.361 g of solid VCl_3 to 90 ml of 0.1 M $Fe(NO_3)_3$ solution in a 2 L plastic bottle, precipitate V-containing ferrihydrite with 162.0; 165.4; 170.8; 175.6; and 186.2 ml resp. of 5 M KOH while purging with N_2 to minimize oxidation of V^{3+} and make up to 1800 ml with dist. water. Hold in an oven at 70 °C for 48 days with occasional shaking. Wash free from electrolytes and freeze or air dry (Schwertmann and Pfab, 1994).

Product Description

These goethites contain up to 6 mol % V. The V-goethites have a definite greenish tinge up to 6 Y (see Plate IV and Fig. 3-7). The dominant wave length moves from 578 to 548 nm as the V content increases. Acicular crystals predominate, but large well developed twins were formed as well (Fig. 5-16). Because the V^{III} cation has, within experimental error, the same size as the Fe^{III} cation (64.0 vs. 64.5 pm), no change in unit cell size was found nor can it be expected. The trivalency of structural V is thus confirmed; further support comes from the absence of an ESP signal for $V^{IV}$and from an equivalent amount of Fe^{2+} ions in an acid digest of V-goethites.

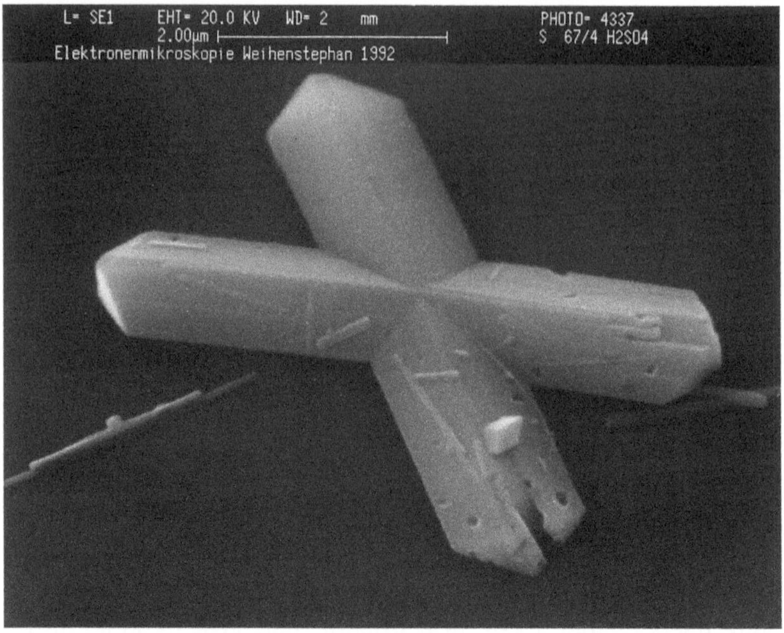

Fig. 5-16. Large goethite twin with well developed {101} side and {210} terminal faces produced from 2-line ferrihydrite in 0.3 M KOH at 70 °C in the presence of 6 mol% V^{III}. Bar = 1 µm. (Schwertmann and Pfab, 1994, with permission).

6 Lepidocrocite

6.1 Introduction

Lepidocrocite is conveniently synthesized by oxidizing a Fe^{2+} containing solution at a pH close to neutral. Maintaining the pH at this level during the entire process is essential and, because protons are produced during the oxidation/hydrolysis process:

$$4\ Fe^{2+} + O_2 + 6\ H_2O \longrightarrow 4\ FeOOH + 8\ H^+$$

buffers are normally used in the older recipes to keep the pH at the desired level (e.g. $Na_2S_2O_3$ by Hahn and Hertrich, 1923; pyridine by Baudisch and Albrecht, 1932; hexamethylene-tetramine by Glemser, 1938; Brauer, 1954; Gehring and Hofmeister, 1994). The recipe described below uses automatic pH control instead of a buffer and thereby avoids any additional chemicals. The type of anion used is also important. Chloride as an inert anion is preferred; carbonate directs the system towards goethite (Carlson and Schwertmann, 1990) and any anion, such as phosphate, with a higher affinity for the Fe oxide surface may completely block crystallization to lepidocrocite.

Lepidocrocite may also form by slow hydrolysis of an acidic Fe^{III} salt solution but the crystallinity is much lower than that of lepidocrocite formed in an Fe^{II} sytem (Schwertmann et al. 2000).

6.2 Preparation

Method

The experimental arrangement consists of a 600 mL glass beaker equipped with a stirrer, a pH electrode, a gas inlet and a burette (containing 1 M NaOH) with the outlet just above the solution (Fig. 6-1). The burette should be connected to a pH-stat unit through a magnetic valve. The gas supply should be delivered through a porous frit located near the bottom of the beaker close to the stirrer to ensure a uniform distribution of gas.

Add 300 mL distilled water to the beaker and dissolve 11.93 g of unoxidized crystals of $FeCl_2 \cdot 4\ H_2O$ (60 mmoL Fe) with stirring. If the crystals have an ochreous tinge, filter the solution to remove any precipitate of akaganeite. Adjust the pH of the acidic solution to pH 6.7–6.9 with NaOH. Open the air supply and adjust the flow rate to 100 mL/min. During oxidation, the color of the suspension changes from dark greenish blue to grey and finally to orange. Protons produced during the oxidation/hydrolysis reaction are constantly neutralized by NaOH added from

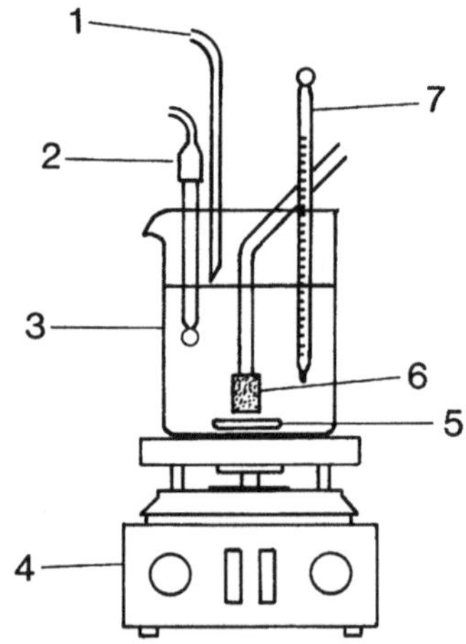

Fig. 6-1. Equipment for lepidocrocite synthesis: 1. Inlet for base addition, 2. Glass electrode; both 1 and 2 are connected to a pH stat instrument, 3. Reaction beaker, 4. Magnetic stirrer, 5. Magnet, 6. Porous ceramic block for air entry, 7. Thermometer.

the burette. After consumption of ca. 120 mL NaOH, oxidation is complete and the pH remains constant without further addition of alkali. The entire reaction takes 2–3 hr. Centrifuge, wash and dry the product.

Product description

This method yields approximately 6 g of reasonably well crystalline lepidocrocite as seen from the X-ray diffractogram and the electron micrograph (Fig. 6-2). The crystals are lath-like and elongated in the *c*-direction. The surface area is around 70–80 m^2/g. An IR spectrum is shown in Fig. 6-2, and a Mössbauer spectrum, recorded at 50 K with and without an external field, in Fig. 6-3.

Comments

Control of pH is crucial for obtaining a pure product. To avoid ferrihydrite, and also a longer oxidation period, the pH should not fall below 5. To avoid magnetite/maghemite, the pH should not exceed 7.5–8. Higher pH values may occur locally at the NaOH inlet if this dips into the suspension and if stirring is not sufficient. Where magnetite forms, it can be recognized by its black color and its adhesion to a magnet held against the reaction vessel. Once formed, magnetite will not disappear on further aeration of the suspension. Although carbonate in the system favors the formation of goethite, the CO_2 content of the purge air may be tolerated. Lowering the pH from 7 to 5 may lower the crystallinity of lepidocrocite (Schwertmann and Thalmann, 1976).

Use of $Fe(ClO_4)_2$ (0.01 M) instead of $FeCl_2$, as above, gives maghemite (Krishnarmurthy and Huang, 1993). When hexamethylene tetramine (urotropin) is used as a buffer, very thin laths contaminated with carbon result. These laths often aggregate into large, spiky aggregates.

The synthesis of *Al-substituted* lepidocrocite needs special precautions because Al may promote goethite in a Fe^{II} system. However, poorly crystalline lepidocrocites with up to 12 mol% Al in the structure have been produced by oxidation of mixed $FeCl_2/Al(NO_3)_3$ solutions with CO_2-free air at pH 8 in a NH_3/NH_4Cl-buffer (0.2 M NH_3 + 0.2 M NH_4Cl; 1:19) with the pH being kept constant by adding M NH_3 dropwise (Schwertmann and Wolska, 1990; Wolska et al. 1994). Note that Al can only be incorporated into the lepidocrocite structure, if the pH of the system is close to 8.

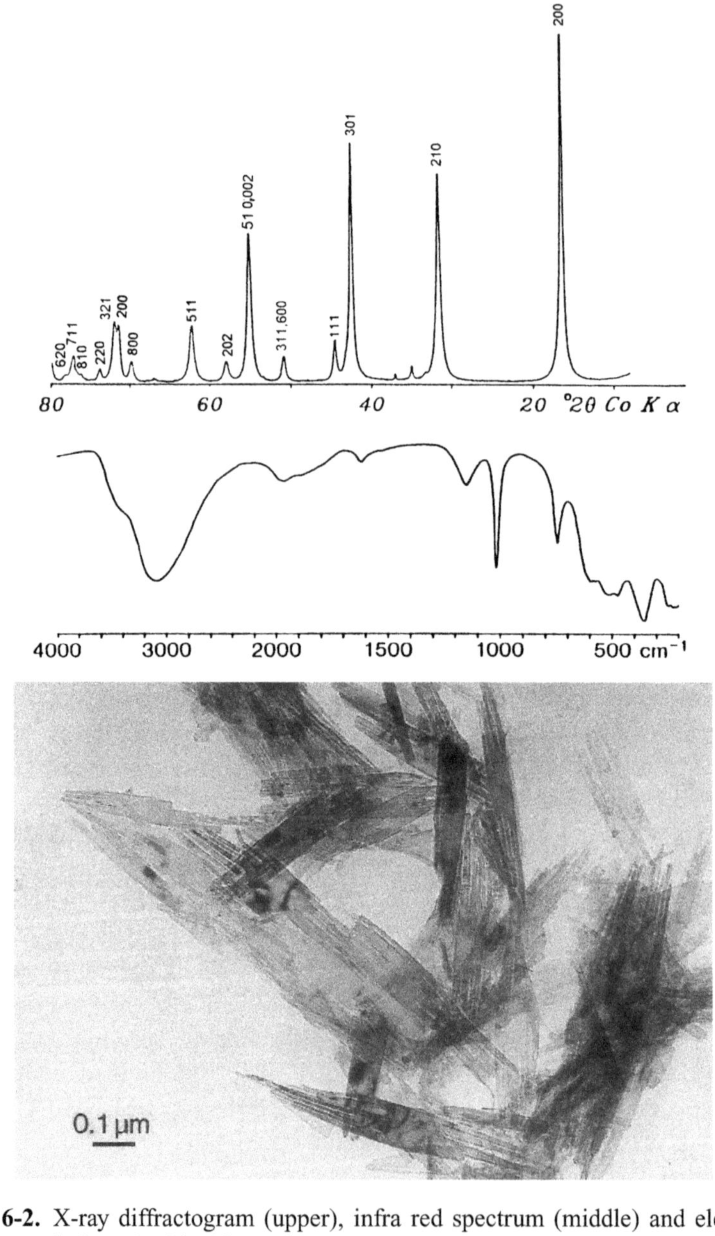

Fig. 6-2. X-ray diffractogram (upper), infra red spectrum (middle) and electron micrograph (lower) of lepidocrocite.

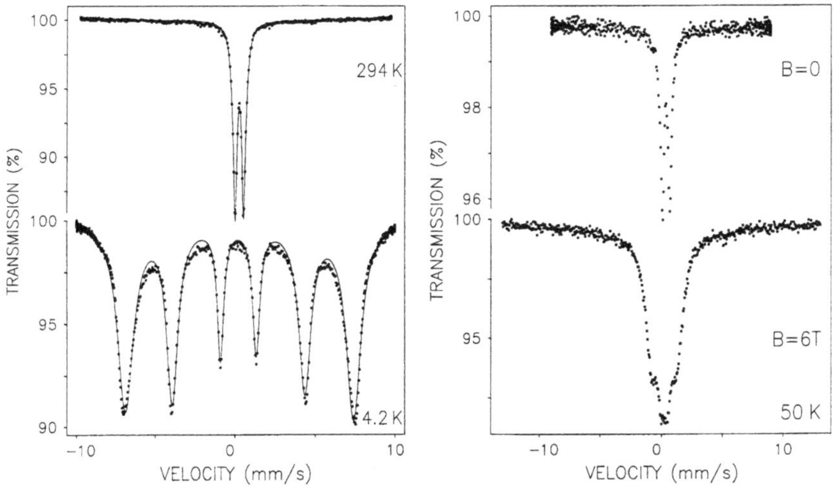

Fig. 6-3. Mössbauer spectra of lepidocrocite. Left, spectra recorded at 294 K and 4.2 K. Right, spectra recorded at 50 K without and with an external field of 6 T. Left, Courtesy E. Murad. Right, Murad (1996) with permission.

6.3 Other Methods

Monodomainic lath-like crystals up to 0.5–1 μm long were synthesized by air oxidation (0.5 L/min) of a 0.2 M $FeCl_2$ solution in a NH_3/NH_4Cl-buffer at 50 °C and pH 6 (see Fig. 3-11) (Giovanoli and Brütsch, 1974). The pH was held constant by adding 0.1 M NH_3 from a burette. The sample surface area was 15 m^2/g.

Tabular lepidocrocite has been obtained by holding metallic iron particles in 0.25 M HNO_3 at 70 °C and aerating for 6 hr at 1.5 L/hr (Kiyama et al. 1972).

7 Feroxyhyte

7.1 Introduction

Feroxyhyte is a poorly ordered form of δ-FeOOH (and therefore called δ′-FeOOH) discovered and introduced as a mineral by Chukhrov et al. in 1977. It is structurally related to the ferrimagnetic δ-FeOOH described by Glemser and Gwinner in 1939.

7.2 Method

Prepare a 0.1 M Fe^{II} solution from unoxidized crystals of $FeCl_2 \cdot 4\,H_2O$. Filter the solution to remove any ochreous precipitate. Place a 1 L glass beaker on a magnetic stirrer and install a pH electrode. Pour 300 ml of the $FeCl_2$ solution into the glass beaker. Raise the pH of the solution from the initial ca. 3 to 8 by adding 5 M NaOH dropwise with constant stirring. A green precipitate (green rust) will form. Rapidly pour in 40 ml of H_2O_2 (30%). The precipitate turns reddish brown within a few seconds and the pH drops to 2–2.5 owing to release of protons (p. 62). The reaction is violent and should be carried out in a fume hood for safety reasons. To improve flocculation raise the pH back to 8 (close to the zero point of charge) by adding 5 M NaOH dropwise with stirring. Centrifuge, wash and dry.

Product description
This method yields about 2.5 g of feroxyhyte with broadened X-ray lines (Fig. 7-1) whose widths follow the order 100, 110 < 101 < 102 indicating

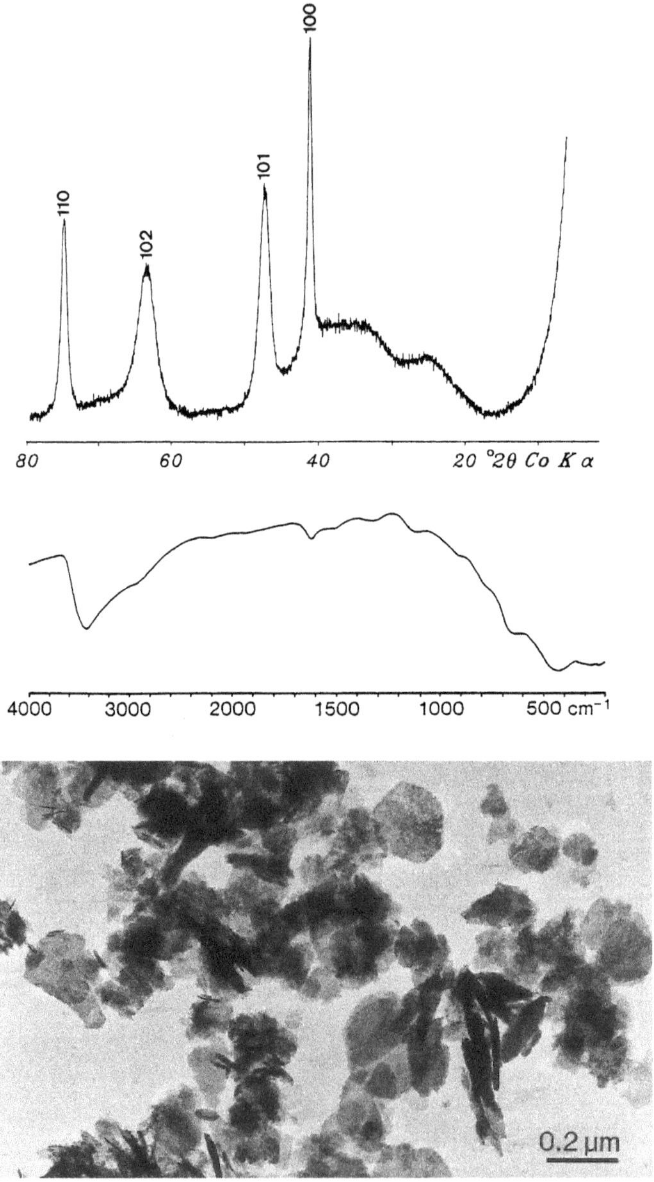

Fig. 7-1. X-ray diffraction pattern, IR spectrum and electron micrograph of feroxyhyte.

much better crystal growth in the *a* direction than in the *c*-direction. This is also confirmed by the morphology of the crystals which form thin plates (Fig. 7-1). The surface area is 200 m^2/g. Figure 7-1 also shows the IR spectrum of the product. A detailed structural analysis was carried out by Drits et al. (1993).

Comments

An initial pH of 7–8 before addition of H_2O_2 is probably the optimum value for the production of reasonably well crystalline material. Lowering the pH yields less crystalline feroxyhyte.

Very rapid oxidation is essential for formation of feroxyhyte. As the oxidation rate is lowered, lepidocrocite and/or magnetite may form.

Koch et al. (1985) produced feroxyhyte in the same way as described above, except that the 0.1 M $FeCl_2$ solution was produced by dissolving Ferrum reductum (a mixture of iron metal and Fe_3O_4) in HCl. They obtained platy crystals 15–90 nm across and 2–3 nm thick with a surface area of 110 m^2/g and a magnetic hyperfine field at 5 K of 52.5 T. The magnetic properties of this material are described in Bender-Koch et al. (1995).

Well crystalline, ferrimagnetic δ-FeOOH can be produced by H_2O_2 oxidation of $Fe(OH)_2$ at high pH (>14) (Feitknecht, 1959).

8 Ferrihydrite

8.1 Introduction

Ferrihydrite is generally the initial precipitate that results from rapid hydrolysis of Fe^{III} solutions. Its crystallinity, i.e. crystal size and order, is usually lower than that of any of the other Fe oxides described except feroxyhyte and schwertmannite. It is usually named according to the number of its XRD peaks, with 6–8 broad peaks for "well" crystalline (6-line-) ferrihydrite and only two very broad ones for the most poorly crystalline form (2-line-ferrihydrite). The 2-line ferrihydrite is commonly but incorrectly called "hydrous ferric oxide (HFO)" or , "amorphous iron oxide". In natural environments all forms of ferrihydrite are widespread usually as young Fe oxides and they play an important role as an active sorbent due to their very high surface area.

The crystallinity of ferrihydrite is determined by the rate of hydrolysis and/or the presence of interfering compounds. Rapid, forced hydrolysis of Fe^{III} salt solutions under very acidic conditions at elevated temperatures (e.g. 80 °C) for a short period of time leads to 6-line ferrihydrite whereas rapid hydrolysis at RT and close to neutral pH produces 2-line ferrihydrite (Chukhrov et al., 1973; Schwertmann and Fischer, 1973). The 2-line material transforms to goethite and/or hematite if stored under water even at ambient temperature and is therefore often used in the laboratory as the precursor of goethite and hematite. Even in the air-dry state, transformation to hematite and goethite has been observed after several years (Schwertmann et al. 1999). The transformation can be effectively retarded or even completely blocked by small amounts of sorbates such as silicate, phosphate, a range of organics and also by coprecipitated Al (Schwertmann and Fischer, 1973; Karim, 1984; Cornell et al. 1987; Schwertmann, 2000 a).

A full range of ferrihydrites with between 2 and 6–8 XRD lines can be obtained either by varying the rate of hydrolysis of Fe^{III} salt solutions at RT or, alternatively, by oxidizing a Fe^{II} salt solution in the presence of various concentrations of dissolved silicate. These methods will be described in the following chapter.

Two-line ferrihydrite, for use as a catalyst, can also be synthesized by thermal decomposition of iron penta carbonyl, $Fe(CO)_5$, in a stream of moist air at 500 °C (Kosowski, 1993; Zhao et al.1993). It is a free flowing, reddish brown powder with a much lower bulk density than freeze-dried ferrihydrite prepared as above, probably because of much weaker aggregation. Small changes in the reaction conditions (water content of the air, duration of heating) may induce hematite formation. An analogous recipe involved slow thermal decomposition of trinuclear aceto-hydroxy Fe^{III}-nitrate for 20–40 hr in air in an attempt to simulate the red pigment on the Martian surface (Morris et al. 1991).

In a Fe^{III} solution containing some Al, the latter coprecipitates at a lower pH than expected for pure Al solutions. Taylor (1988) has called this "induced hydrolysis" of Al. No significant XRD-peak shift indicative of structural incorporation of Al was, however, discernable. This may be due to the poor crystallinity of the Al containing ferrihydrite.

Two-line ferrihydrites (or related compounds) containing substantial amounts of Si, C, As, and U have been found in surface environments and some shift of the XRD peaks has been observed. This finding indicates the ability of ferrihydrite to retain these elements and thereby to exert a purifying effect on natural waters.

An extensive review on ferrihydrite has been puslished by Jambor and Dutrizac (1998).

8.2 6-Line Ferrihydrite

Method

Preheat 2 L of distilled water to 75 °C in an oven, withdraw the water and add 20 g unhydrolysed crystals of $Fe(NO_3)_3 \cdot 9\ H_2O$ with rapid stirring. Return to the oven and leave there for a further 10–12 minutes.

During this time the solution changes from gold to dark reddish brown indicating the formation of Fe hydroxy-polymers. No precipitate should form. Cool rapidly by plunging into ice water, transfer to a dialysis bag and dialyse for at least three days, changing the water several times each day. Collect the product and freeze dry it.

Product Description

This method gives around 5 g ferrihydrite with a BET surface area of 200–300 m^2/g. The X-ray diffraction pattern usually shows six broadened lines (Fig. 8-1) and with the electron microscope, subrounded particles of 2–6 nm in diameter are seen (Fig. 8-2). At high magnification (HRTEM) the crystalline nature of the product is obvious (Fig. 8-3) (Eggleton and Fitzpatrick, 1988; Janney et al. 2000). The product may contain a trace of nitrate (IR absorption band at 1380 cm^{-1}) even after prolonged dialysis against water (Fig. 8-1). Mössbauer spectroscopy (Fig. 8-4) shows magnetic ordering at 4K but the spectrum has broad resonance lines and is best fitted to a hyperfine field distribution (Fig. 8-4) (Murad and Schwertmann, 1980; Murad et al., 1988).

Comments

The length of time of hydrolysis is critical; with longer hydrolysis times goethite and hematite will form. It is essential that the Fe^{III}-nitrate is added to the preheated water because goethite may form during the heating phase, if the $Fe(NO_3)_3$ is dissolved in water at ambient temperature.

8.3 2-Line Ferrihydrite

Method

Dissolve 40 g $Fe(NO_3) \cdot 9\ H_2O$ in 500 ml distilled water and add 330 mL M KOH to bring the pH to 7–8. The last 20 ml should be added dropwise with constant checking of the pH. Stir vigorously, centrifuge and then dialyse rapidly until free from electrolytes. Freeze dry and store as a solid.

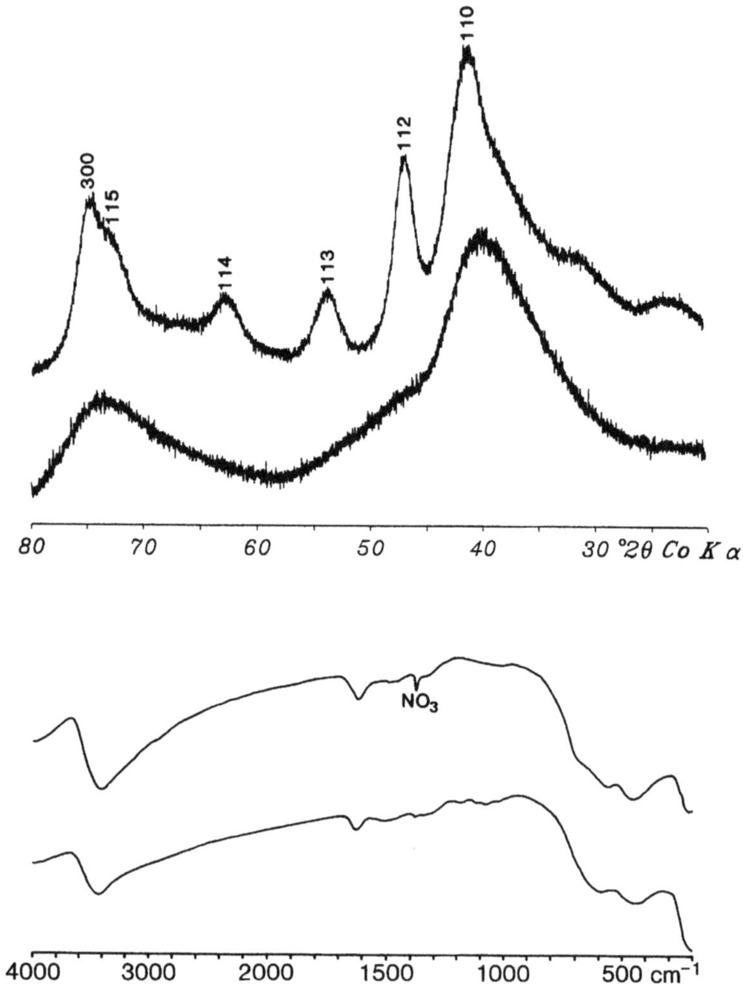

Fig. 8-1. X-ray diffractograms and IR spectra of a 6-line ferrihydrite (upper curves) and a 2-line ferrihydrite (lower cuves).

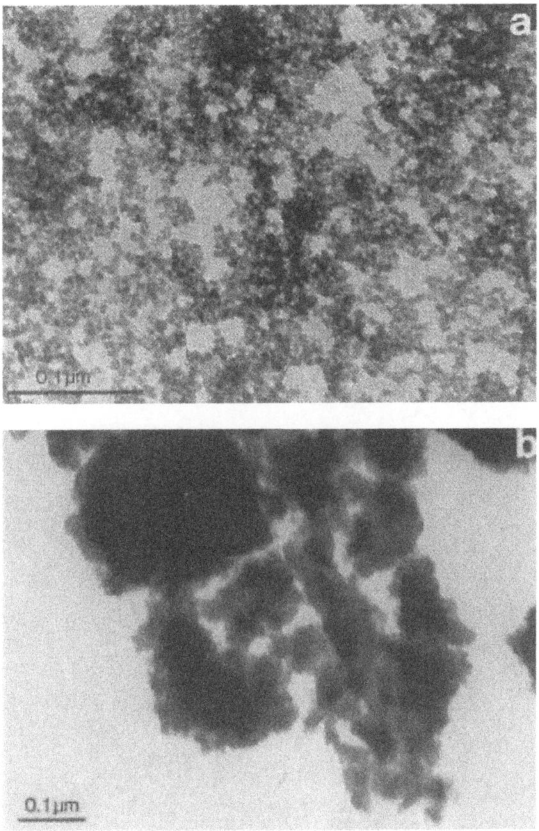

Fig 8-2. Electron micrograph of a 6-line (a) and a 2-line (b) ferrihydrite. (Cornell and Schwertmann, 1996; with permission.)

Product Description

This method gives 10 g 2-line ferrihydrite. X-ray diffraction shows two very broad peaks at 0.24 and 0.15 nm (Fig. 8-1). The IR spectrum is almost featureless in the range above 1000 cm^{-1}. With high resolution TEM (HRTEM), clear evidence of some structural order, in agreement with XRD, can be seen (Fig. 8-3). The existence of a continuous transition between 2-line and 6-line ferrihydrite (see 8.3) indicates the close relationship between 2- and 6-line ferrihydrite and justifies the name 'ferrihydrite' for the 2-line product.

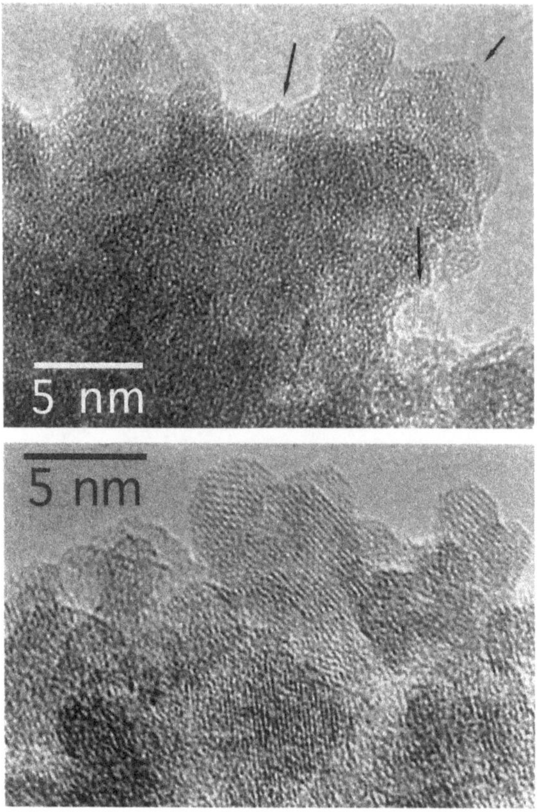

Fig. 8-3. High-resolution electron micrograph of a 2-line (upper) and a 6-line (lower) ferrihydrite. The micrographs show lattice fringes and possibly hexagonal crystal shapes, both being better expressed in the 6-line than in the 2-line form (Courtesy D.E. Janney), (Janney et al. 2000; with permission).

The BET-surface area of freeze-dried products lies in the range of 200–320 m^2/g. Assuming spherical particles with a diameter of 3 nm and a density of 4 g/cm^3, a surface area of ca. 600 m^2/g results. The large discrepancy between this value and the BET surface may be due to the marked aggregation of the nano particles (Fig. 8-2), which makes part of the surface inaccessible to the sorbate (N_2). Freeze- or air-drying aggravate this effect. Once ferrihydrite is aggregated by drying, it is very difficult to redisperse. Ferrihydrite may, however, be held as a stable sus-

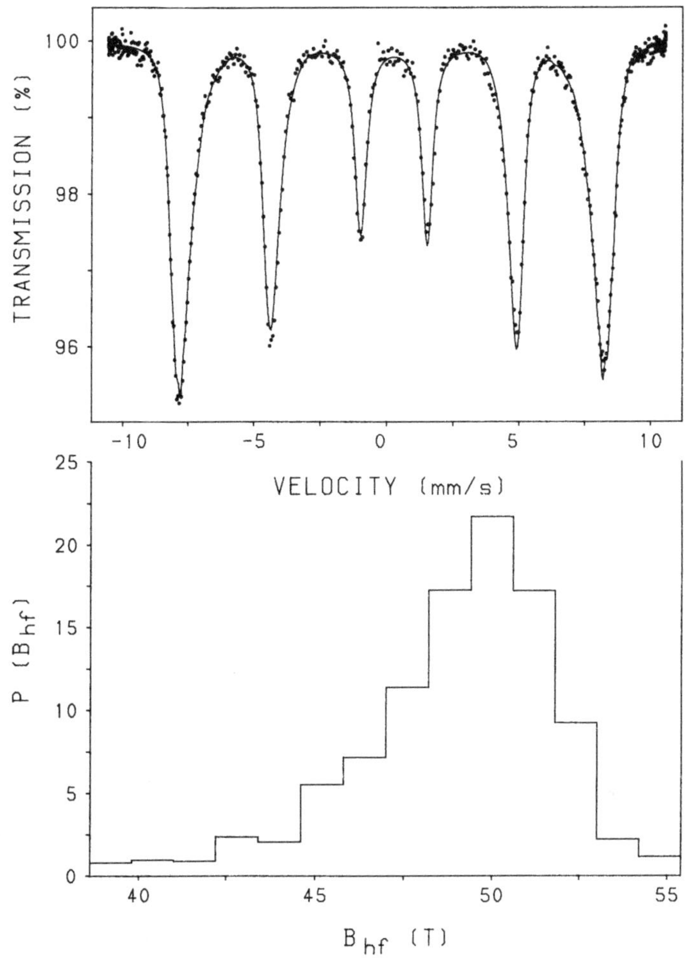

Fig. 8-4. Mössbauer spectrum of 6-line ferrihyrdrite recorded at 4.2 K (upper) and its hyperfine field distribution (lower). (Courtesy E. Murad)

pension after flocculation in an aqueous system if the surface charge of the particles is increased, preferably by lowering the pH to ca. <4, i.e. moving away from the PZC. On the other hand, as the pH is lowered and the particles are dispersed, goethite precipitates from such suspensions, although very slowly (Glasauer et al., 1999). Therefore, ferrihydrite must be used for adsorption, dissolution and similar studies, soon after preparation.

Comments

To avoid local pH values >8, which may induce some goethite formation, vigorous stirring is essential. Equally important is the rapid removal of impurities by centrifugation/washing followed by dialysis for 2–3 days. Raven et al.(1998) observed a trace of goethite after storing a ca. 10 g/L ferrihydrite suspension for >4 weeks in 0.1 molar NaCl solution at pH 7.5 even at 2 °C. The degree of transformation of ferrihydrite to goethite/hematite at pH 4, 5, 6, and 7 and 25 °C in an aqueous suspension (0.125 M Fe) as indicated by the oxalate-soluble proportion was 66; 72; 77; and 83%, respectively, after 1 year and followed 1st-order kinetics (Schwertmann et al. 2000). Drying drastically retards the rate of transformation.

8.4 Ferrihydrites with a Range of Crystallinities

(acc. to D. G. Lewis)

Method

A set of ferrihydrites with a range of crystallinities and between 2- and 6+-X-ray diffraction peaks can be produced at RT either by varying the rate of hydrolysis of a $Fe(NO)_3$ solution or by oxidizing a $FeCl_2$ solution containing different levels of soluble silicate.

The reaction vessel shown in Fig. 6-1 can be used to control pH, reagent addition and gas flow. The titration is carried out at RT and with vigorous stirring.

Rate-of-hydrolysis Series

Place 50 mL of a 0.1 M $Fe(NO)_3$ solution (5 mmol Fe) into the reaction vessel and automatically titrate it at a fixed rate of 1ml/min with 15 mmol NaOH to a final pH of 7. To vary the rate of hydrolysis, the NaOH concentration is varied between $7.7 \cdot 10^{-4}$ and 1 M in 8 steps hence the time for total neutralization is between 0.25 and 130 h and the corresponding rate of OH addition (hydrolysis rate) between 0.38 and

200 μmol OH/mmol Fe/min. After complete hydrolysis (OH/Fe = 3), the solids are obtained by centrifugation, washed with 70mL of distilled water, dried at 40 °C and gently ground.

Silicate Series

Prepare a set of 0.3M NaOH solutions containing 6.6–73 mmol/L Si in 5–8 steps using a Na-silicate solution. Add 50 or 100 ml of a filtered fresh 0.1 M $FeCl_2$ solution to the open reaction vessel (Fig. 6-1). Vigorous stirring is sufficient to oxidize the $FeCl_2$ solution. Keep the pH constant at 6.5 by adding the Si-containing NaOH solutions. The NaOH is kept CO_2-free with a trap consisting of a mixture of solid NaOH, $Ca(OH)_2$ and CaO. After complete oxidation (no further base consumption), separate and wash the solids by centrifugation.

Product Description

The method gives ca. 0.5 g (1 g) ferrihydrite. Scaling up has not been tried yet. Fig. 8-5 shows XRDs of the hydrolysis and the Si series. In the hydrolysis series, goethite and/or lepidocrocite, i.e. the common FeOOH forms, appear when the hydrolysis rate falls below ca. 17 μmol OH/min and in the Si series as the Si concentration falls below 10 μmol/L.

Whereas the hydrolysis products are chemically pure, those made in the presence of Si contain between 10 and 75 mg Si/kg. The surface area ranged from between 210–240 and 175–350 m^2/g for the two series, respectively. The hyperfine properties (Mössbauer spectra) vary greatly. Decreasing crystallinity is coupled with decreasing magnetic hyperfine fields at 4 K and with a decreasing magnetic blocking temperature (temperature at which magnetic ordering sets in) (Schwertmann et al. 2000 a).

Comments

The silicate has to be combined with the base (to neutralize the protons during oxidation) in order to guarantee a continuous supply of Si while the ferrihydrite is formed. This process simulates ferrihydrite formation in many natural environments. If, however, Si is added at the beginning of oxidation in one lot, the dissolved silicate is completely adsorbed by the first ferrihydrite to form and well crystalline lepidocrocite will form from the remaining Fe^{II}, because dissolved silicate is no longer present.

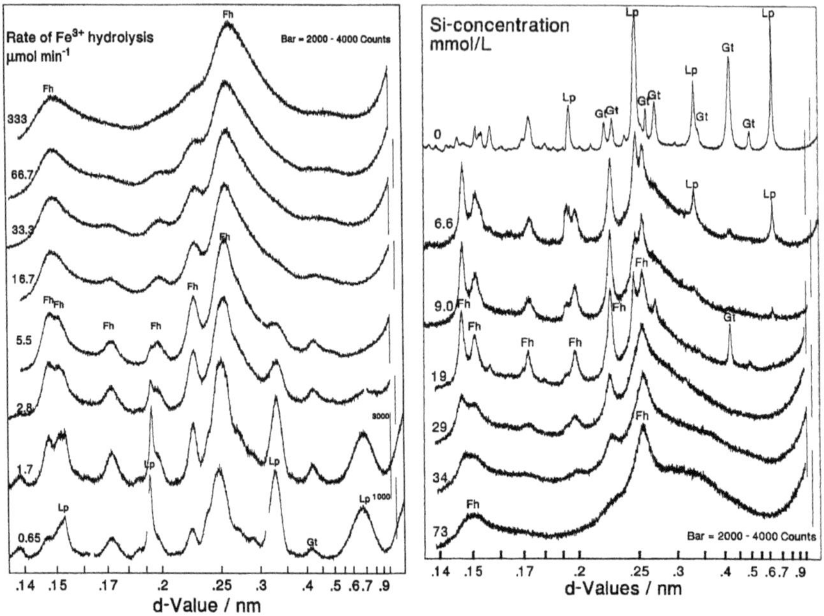

Fig. 8-5. X-ray diffractograms of ferrihydrites varying in crystallinity obtained by hydrolysing a $Fe(NO_3)_3$ solution at different rates (left) and by oxidizing $FeCl_2$ solutions with different silicate concentrations (right). Fh: ferrihydrite; Gt: goethite; Lp: lepidocrocite.

9 Akaganéite

9.1 Introduction

Akaganéite is prepared by hydrolysis of a Fe^{III} chloride solution. The halogen ion (together with H_2O molecules) occupies the 0.5×0.5 nm^2 interstices in the tunnels in the structure and appears to direct this structure and to stabilize it. The chloride content of akaganeite varies between 1 and 7%. It can be reduced by thorough washing or dialysis, but it seems impossible to extract all the chloride without collapse of the structure and, depending on the temperature, transformation to goethite or hematite.

Akaganeite cannot be prepared at pHs above 5 because the OH^- ion is far more competitive than the chloride ion for structural sites. Akaganeite displays two morphologies: somatoids or cigar shaped crystals and much smaller rod-like crystals. Samples of somatoids frequently contain a proportion of twinned crystals, whereas the rod-like crystals are never twinned.

9.2 Preparation by Hydrolysis of Acidic $FeCl_3$ Solutions (Somatoids)

Method

Dissolve 54.06 g $FeCl_3 \cdot 6\,H_2O$ in 2 L twice distilled water (i.e. 0.1 M solution). Hold in a 3 L closed polyethylene or glass flask at 40 °C for 8 days. During this period the bright, gold solution becomes a lighter yellow and a compact, yellow precipitate forms. The pH drops from 1.7 to around 1.2.

Product Description

The method gives 5 g akaganeite (27% yield). The crystals are well-formed somatoids around 300 nm in length and elongated in the c-direction (Fig. 9-1). The sample has a surface area of around 30 m^2/g and sharp XRD peaks (Fig. 9-2), indicating a well-crystallized material. An IR spectrum is shown in Fig. 9-2. The Mössbauer spectrum shows three doublets at 293 K and three sextets in the magnetically ordered state at 78 K (Fig. 9-3).

Comments

The drop in pH during hydrolysis of Fe^{III} chloride solution upon heating leads to incomplete precipitation of akaganeite: only 20–30% of the total Fe in the system is precipitated over the range 40–70 °C. A higher yield may be obtained by increasing the temperature or by twofold dilution of the solution (50% yield). Dilution of the system does not affect the crystallinity of the product, whereas increasing the temperature has the disadvantage that it leads to poorly crystalline material most probably because of more rapid crystallization.

Hematite forms competitively with akaganeite at reaction temperatures above 90 °C and goethite forms competitively if seed crystals of goethite are added (Atkinson et al., 1977).

Akaganeite may transform into either goethite or hematite at high enough temperature if left in the mother liquor long enough. However, at 70 °C neither transformation nor any Ostwald* ripening was observed after a 5 month period.

Akaganeite may also be precipitated from Fe^{III} fluoride solutions (Bernal et al. 1959; Naono et al. 1993), but not from bromide or iodide solutions; the latter ions are too large to fit into the tunnels in the structure.

* Ostwald ripening is the growth of large particles in a suspension at the expense of the smaller ones which redissolve.

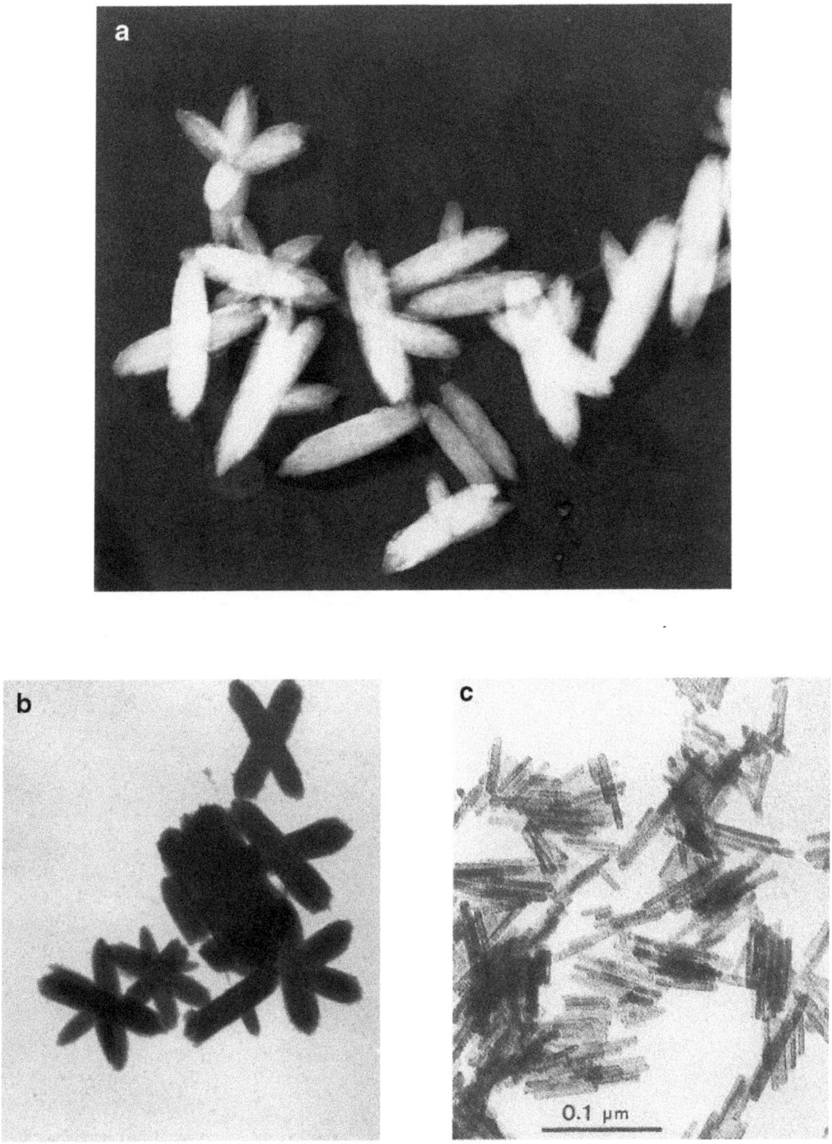

Fig. 9-1. Electron micrographs of (a) somatoidal, (b) rod-like akaganéite, (c) Si-akaganéite. Bar = 0.1 µm.

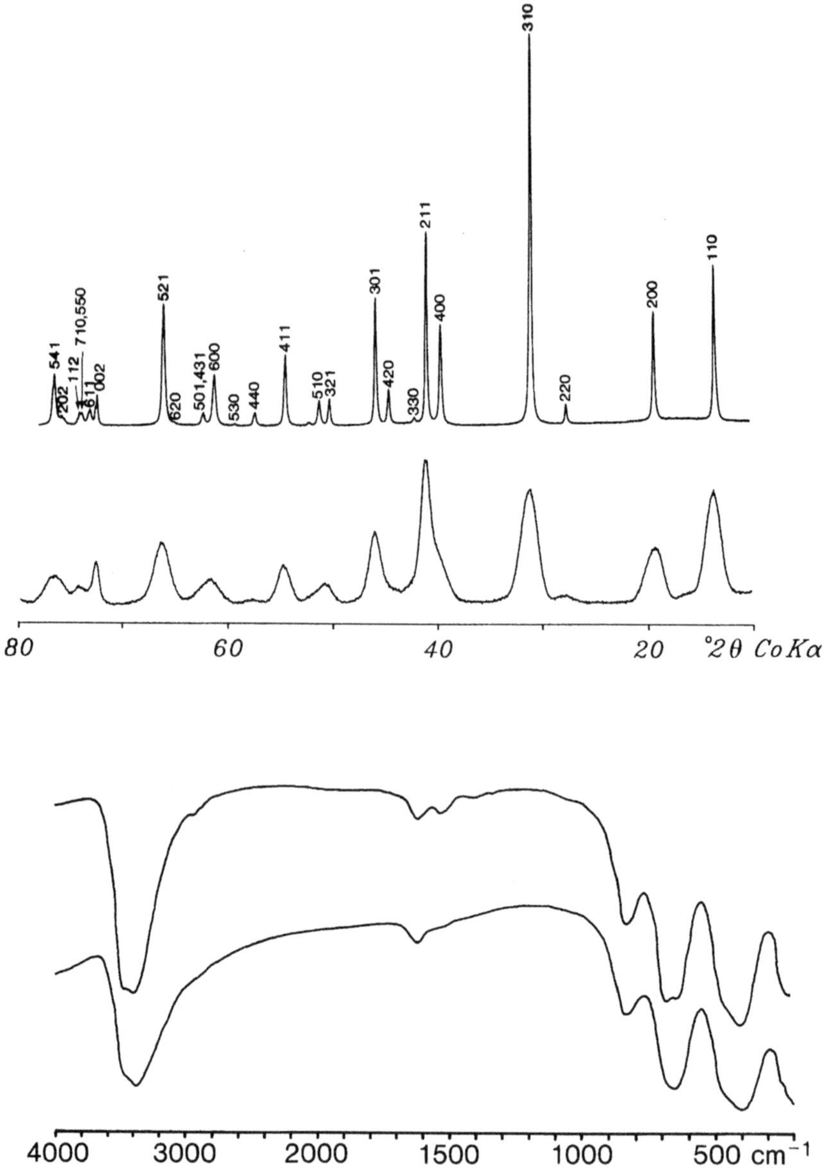

Fig. 9-2. X-ray diffractogram and IR spectrum of somatoidal (upper) and rod-like (lower) akaganeite.

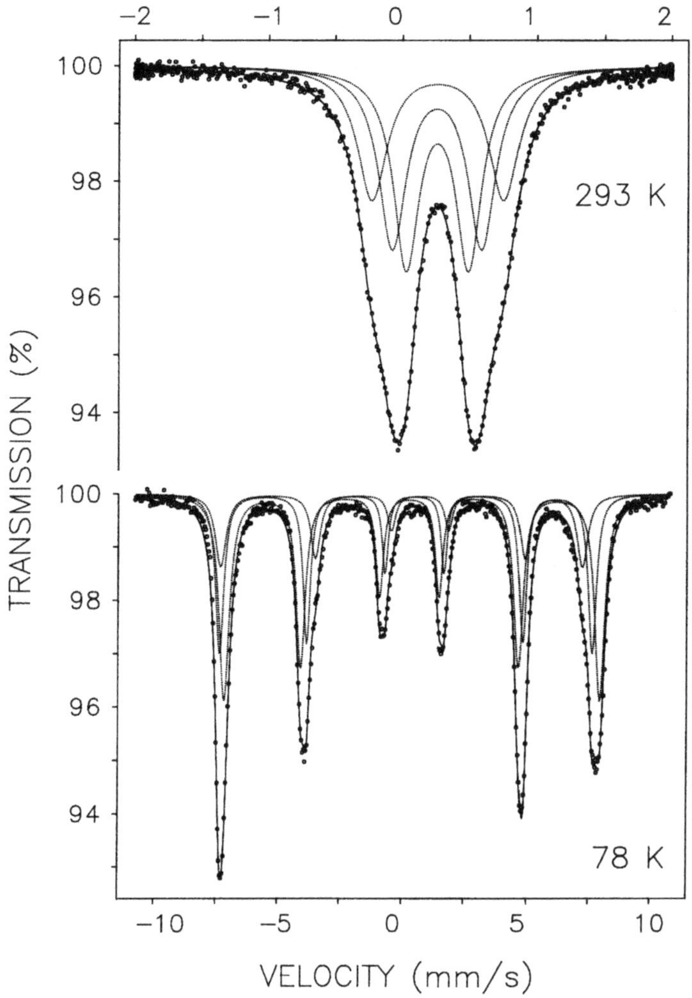

Fig. 9-3. Mössbauer spectra of somatoidal akaganeite recorded at room temperature and at 78 K (see Murad, 1979).

9.3 Preparation by Hydrolysis of a Partially Neutralized $FeCl_3$ Solution (Rod-like Crystals)

Method

Dissolve 27.3 g $FeCl_3 \cdot 6H_2O$ in 100 mL twice distilled water (1 M solution) and add 75 mL 1 M NaOH solution for partial neutralization (OH/Fe = 0.75). A temporary, brown precipitate forms, but redissolves on shaking to give a clear, dark brown solution. Allow this solution to stand at room temperature for 50 hours and then add 20 mL 10 M NaOH solution and heat the resulting suspension at 70 °C in a closed polyethylene flask for eight days. Wash and dry the brownish-yellow precipitate. Before washing akaganeite by dialysis, the supernatant liquid must be removed by centrifugation. Otherwise the Fe^{3+} remaining in the supernatant liquid will precipitate as ferrihydrite owing to a rise in the pH of the system during dialysis.

Product Description

The method gives 6 g akaganéite (70% yield) with a surface area of around 100 m^2/g. The crystals are narrow rods up to 50 nm in length (Fig. 9-1b). The XRD pattern shows broad peaks (Fig. 9-2) resulting from the small size of the crystals. An IR spectrum is shown in Fig. 9-2.

9.4 Si-containing Akaganéite

Method

Dissolve 40.4 g $Fe(NO_3)_3 \cdot 9\,H_2O$ in 1 L twice distlled water and add sufficient Si solution (Titrisol, Merck: $SiCl_4$ with 1000 mg Si/L in 14% NaOH) to give a Si/Fe ratio of 0.07. Hold the solution in a 1 L polyethylene bottle in a 70 °C oven for 24 hr. Wash the yellow precipitate and dry at 50 °C.

Product Description

The method produces 2.25 g akaganéite with a surface area of 30–40 m^2/g. The crystals are approximately somatoidal and between 200–400 nm long.

This material contains 2.3 mol% Si. Si is released congruently with Fe upon dissolution and is, therefore, considered to be homogenously distributed within the akaganeite crystals, probably in the tunnels (Cornell, 1992).

Comments

The Cl^- ion which is necessary for the formation of akaganeite comes from the Si solution. If less than 0.3 M Cl^- is present, a mixture of akaganeite and goethite forms. The Si content can be increased to 4.5 mol% by using a Si/Fe ratio of 0.14. With higher ratios, the reaction is retarded and the clear brown solution is stable for weeks. Twinning increases as either Si/Fe increases or as the hydrolysis temperature is raised. Si-akaganeite grown at 120 °C (3 hr reaction) is 100% twinned (no hematite forms in this period). Only a small proportion of Si is retained by the solid phase; most remains in the supernatant solution. This may be because the tetrahedrally coordinated Si, if located in the tunnel, must share O atoms with the Fe(O,OH) octahedra and this is difficult to achieve for the SiO_4 tetrahedron.

There was no effect of silicate (0.3–1 mol%) on the crystal size, probably because the silicate molecule is uncharged at low pH.

Increasing the phosphate concentration (up to 2 mol%) reduced the spindle length from ~500 nm to 100–200 nm (Kandori et al. 1992).

10 Hematite

10.1 Introduction

Hematite can be prepared by a variety of routes. The most convenient and common ones are

(1) by thermal dehydration of a crystalline iron oxide hydroxide (e.g. 2 FeOOH $\rightarrow$ $Fe_2O_3 + H_2O$) or an iron salt,
(2) by forced hydrolysis of acidic Fe^{III} solutions and
(3) by transformation of ferrihydrite in aqueous suspension.

Despite the simplicity of the method there are three reasons why *thermal dehydration* recipes are not given here; (1) Hematite produced in this way does not consist of idiomorphic* crystals: at temperatures of up to ca. 500–600 °C pseudomorphs of the precursor result and above 600 °C coalesced crystals with ill-defined crystal faces are obtained and the surface area is low, (2) In natural environments hematite rarely forms at such high temperatures, and (3) natural hematites frequently consist of single, more or less idiomorphic crystals.

Forced hydrolysis of Fe^{III} solutions involves hydrolysing $Fe(NO_3)_3$, $Fe(ClO_4)_3$ or $FeCl_3$ solutions at a temperature close to 100 °C under strongly acidic conditions (pH 1–2). It is believed that, if hematite is prepared from $Fe(NO_3)_3$ or $Fe(ClO_4)_3$, it forms from Fe^{III} hydroxy species (Johnston and Lewis, 1983, 1986), whereas, if hematite is prepared from $FeCl_3$, akaganeite may be an intermediate product which then transforms to hematite via solution (Hamada and Matijevic, 1981).

---

* Idiomorphic crystals are crystals whose crystal habit is determined by internal structure.

In methods 1–3 of section 10.2, parameters such as type of anion, acidity and [Fe] are varied to give monodispersed hematites with idiomorphic crystals of different sizes and shapes.

Hematite production by *transformation of ferrihydrite* (methods 4 and 5) starts with the precipitation of 2-line ferrihydrite which is then converted into hematite in aqueous suspension by a short-range crystallization process within the ferrihydrite aggregates.

This transformation takes place under weakly acid to weakly alkaline conditions and requires the presence of some water within the ferrihydrite aggregates (see chapter 4) (Feitknecht and Michaelis, 1962; Schwertmann.and Fischer, 1966, Schwertmann et al., 1999).

Although the solubility of ferrihydrite in this pH range is low some dissolution may occur leading to formation of goethite. Such preparations therefore often contain some goethite admixed with the hematite particularly if the pH deviates substantially from neutral. To avoid goethite, it is essential to conduct all operations at the transformation temperature (>80 °C) or to raise the temperature to above 100 °C in a pressure vessel (see Other Methods). Goethite can also be avoided by including certain inorganic e. g. Al (Schwertmann et al. 2000), phosphate (Galves et al. 1999) or organic additives (e.g. oxalate (Fischer and Schwertmann, 1975)) which may, however, contaminate the hematite. Since the pH of this system is much higher than that of the forced hydrolysis system precipitation of Fe is essentially complete and the yield is close to 100%.

10.2 Preparation by Forced Hydrolysis of Fe^{III} Salt Solutions

Method 1

Heat 2 L of 0.002 M HNO_3 in an Erlenmeyer or Duran flask to 98 °C in an oven. Once this temperature has been reached, remove the reaction vessel from the oven and add 16.16 g unhydrolysed crystals of $Fe(NO_3)_3 \cdot 9\ H_2O$ (0.02 M Fe) with vigorous stirring. Close the flask, return the solution to the centre of the oven immediately and hold for seven days at 98 °C. Centrifuge the compact, bright red sediment, wash and dry it.

Method 2
Heat 500 ml of a 0.2 M $Fe(ClO_4)_3$ solution (51.64 g of $Fe(ClO_4)_3 \cdot 9\ H_2O$) (without addition of acid) at 98 °C for seven days and proceed as in method 1.

Method 3
Heat 2 L of a 0.002 or 0.001 M HCl to 98 °C, add 10.81 g fresh $FeCl_3 \cdot 6\ H_2O$ to give a 0.02 M Fe solution. Hold this solution in a closed vessel at 98 °C for ten days.

Method 4
Add 400 ml of a 1 M $Fe(NO_3)_3$ solution with a burette over 4 hr to 5 L of boiling distilled water and stir continuously. Cool overnight, purify by dialysis, centrifuge and dry.

Product Description
Method 1 produces ca. 3 g of hematite. The crystals are fairly uniform in size (30–50 nm) and diamond shaped (Fig. 10-1 a). The surface area is around 30 m^2/g. An X-ray diffractogram is shown in Fig. 10-2 and an IR spectrum in Fig. 10-3.

Method 2 produces ca. 6 g of diamond shaped crystals 30–50 nm across with a stepped surface (Fig. 10-1 b). The surface area of 80–90 m^2/g is fairly high, probably because of the surface irregularities.

Method 3 produces ca. 3 g of hematite. The sample prepared in $2 \cdot 10^{-3}$ M HCl consists of subrounded crystals between 30–50 nm across (Fig. 10-1 c) with a surface area of around 30 m^2/g. In 10^{-3} M HCl the crystal size is around 150–200 nm (Fig. 10-1 d) and the surface area is only a few m^2/g. The X-ray peaks of the 0.002 M HCl product are somewhat broader than those of the material produced in 10^{-3} M HCl owing to the smaller crystal size. The Mössbauer spectrum at RT (Fig. 10-4) shows a sextet corresponding to a magnetic hyperfine field of 53.3 T.

Method 4 produces a fine, crystalline hematite with unidimensional crystals of 7–10 nm across (Sorum, 1928).

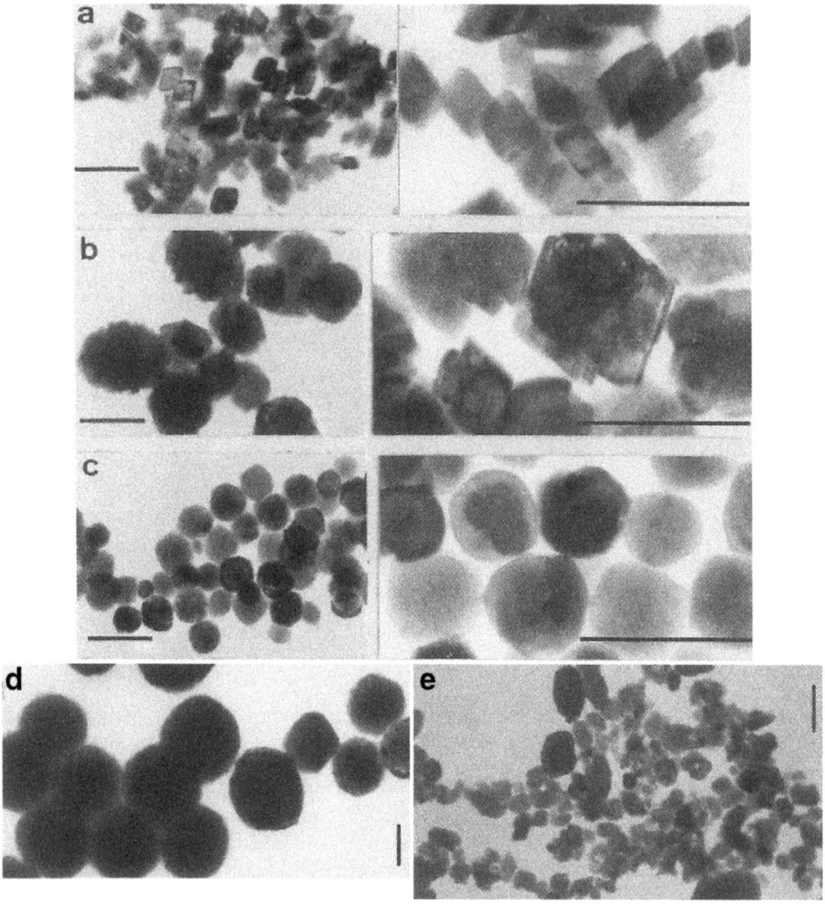

Fig. 10-1. Electron micrographs of hematites produced by forced hydrolysis of Fe^{III}solutions (a–d) and by aging ferrihydrite (e):
a: from $Fe(NO_3)_3$ (Method 1); b: from $Fe(ClO_3)_3$ (Method 2); c, d: from $FeCl_3$ (Method 3), either in 0.002 M HCl (c), or in 1 M HCl (d); e: from 2-line ferrihydrite at pH 8–9 and 90 °C (Method 5).

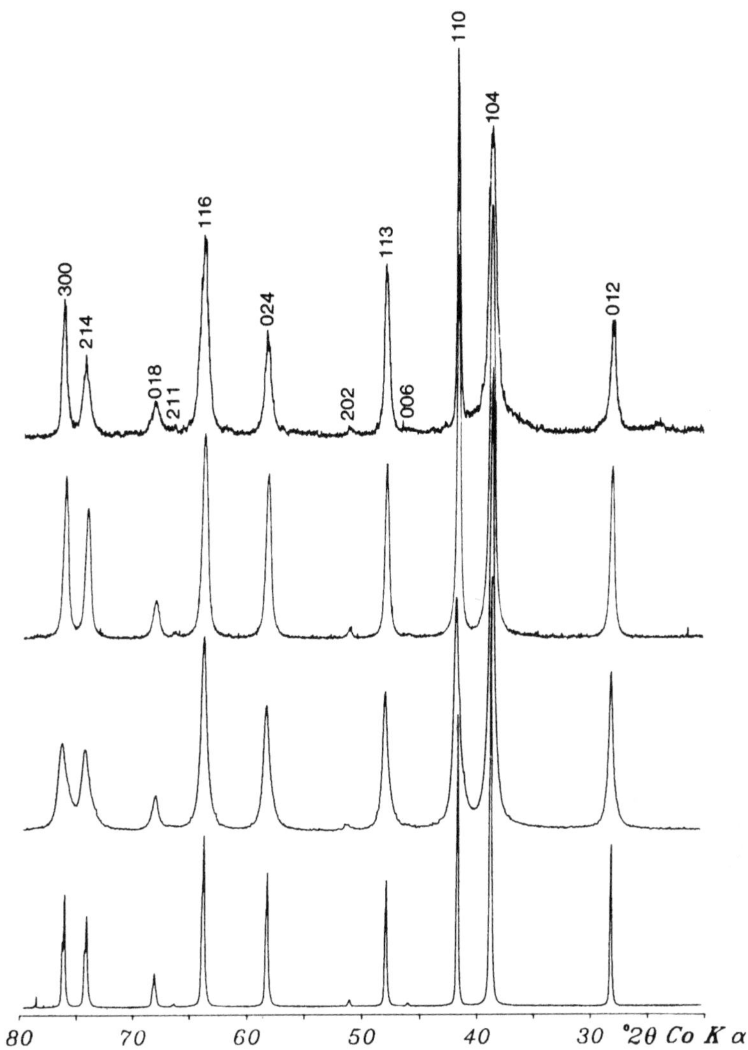

Fig. 10-2. X-ray diffractograms of hematites. From top to bottom: Hematite with 15 mol% Al substitution grown from 2-line ferrihydrite at pH 7 and 80 °C (The peak shift due to Al substitution is too small to be visible on this figure). Hematite from forced hydrolysis of $Fe(NO_3)_3$ (Method 1). Hematite from 2-line ferrihydrite in the presence of oxalate at pH 6.5 (Method 6); Hematite from forced hydrolysis of $FeCl_3$ (Method 3).

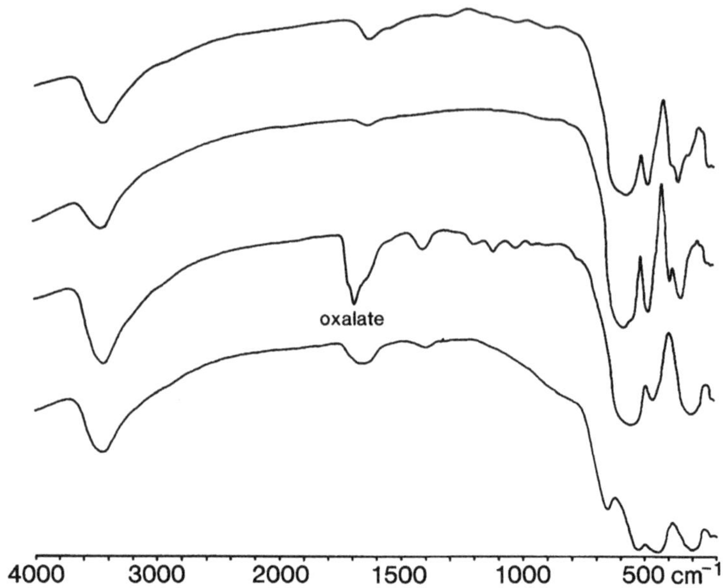

Fig. 10-3. IR spectra of hematites. From top to bottom: Hematite from forced hydrolysis of $Fe(NO_3)_3$ (Method 1); Hematite from 2-line ferrihydrite in the presence of oxalate at pH 6.5 (Method 6); Hematite with 15 mol% Al substitution grown from 2-line ferrihydrite at pH 7 and 80 °C.

10.3 Preparation by Transformation of 2-Line Ferrihydrite

Method 5

Dissolve 40 g $Fe(NO_3)_3 \cdot 9\ H_2O$ in 500 mL twice distilled water preheated to 90 °C and precipitate ferrihydrite with 300 mL M KOH also preheated to 90 °C. Add 50 mL 1 M $NaHCO_3$, preheated to 90 °C, to the brown, voluminous precipitate of ferrihydrite and hold the suspension (pH = 8–8.5) in a closed polyethylene flask at 90 °C for 48 hours.

Method 6

Dissolve 40 g $Fe(NO_3)_3 \cdot 9\ H_2O$ in 500 mL twice distilled water, add 300 mL 1 M KOH, make up the volume to 1 L and add 180 mg of oxalic

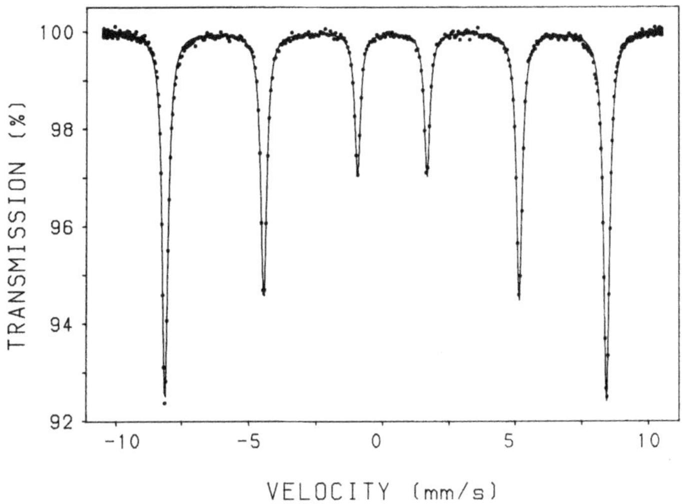

Fig. 10-4. Mössbauer spectrum of hematite recorded at 293 K.

acid (i.e. 0.002 M oxalic acid). Adjust to pH 6.5 with HNO_3 or KOH and heat at 90 °C for 36 hours (Fischer and Schwertmann, 1975). During the transformation the pH falls. To prevent this, the system can be buffered at pH 6.5 with 30 mL 4.4 M imidazole buffer.

Product description

Method 5 gives around 7 g hematite with a surface area of ca. 20–25 m^2/g. The crystals are small platelets (Fig. 10-1 e). The sample has a sharp XRD pattern.

Method 6 produces about 8 g of hematite which consists of grainy-looking ellipsoidal (spindle-type) crystals (Fig. 10-5 d) and has a surface area of ca. 80–90 m^2/g. Again, as seen from the X-ray pattern the sample is reasonably pure and well crystallized (Fig. 10-2) but contains adsorbed oxalate and a trace of goethite (see IR spectrum in Fig. 10-3).

Comments

The mean crystallite dimensions (MCD) along the *a* and *c* directions of the six hematites described above, have been determined by Crosa et al.

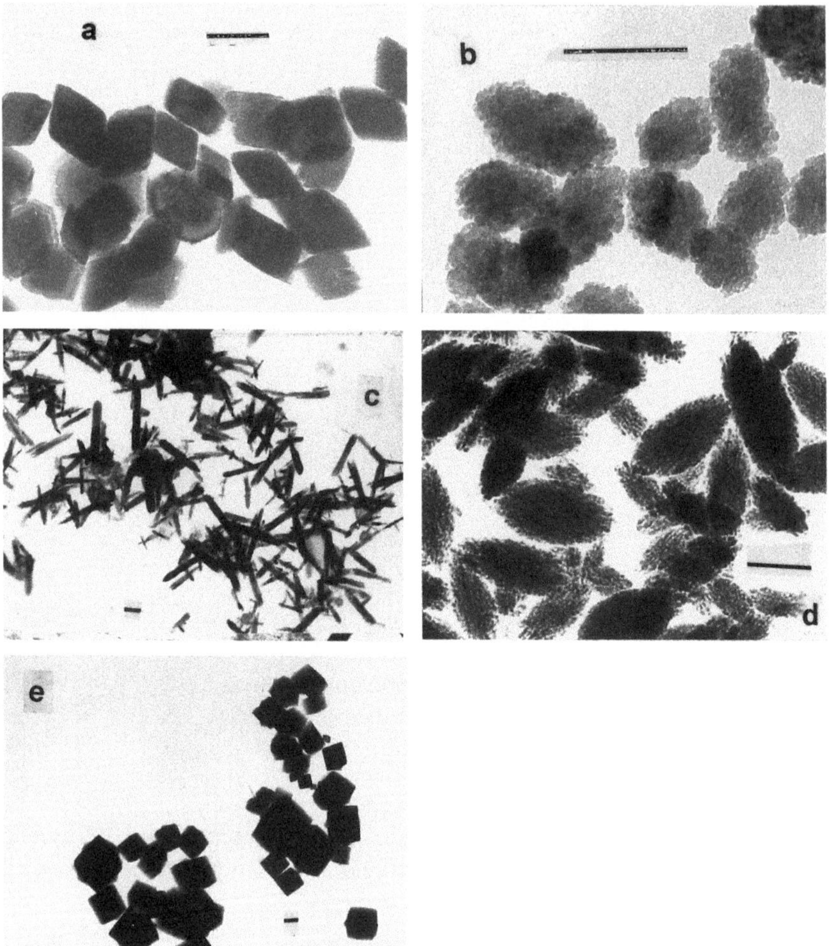

Fig. 10-5. Electron micrographs of hematites produced from ferrihydrite in the presence of various additives: a: in the presence of 0.05 mol% Al at 25 °C and pH 7; b: in the presence of 0.15 mol% Al at 25 °C at pH 4 (Schwertmann et al. 2000; with permission); c: in the presence of 0.0025 M citrate at pH 11 and 80 °C (Schwertmann et al. 1968; with permission); d: in the presence of 0.02 M oxalate at pH 6 and 70 °C (Method 6) (Fischer and Schwertmann, 1975; with permission); e: in the presence of Cu at pH 12.2 and 70 °C (Cornell and Giovanoli, 1993; with permission).

(1999) from XRD line broadening (see Section 3.3). The surface areas calculated from these MCD values correspond quite well to the measured ones. Most of the hematites produced in aqueous systems contain some OH and/or H_2O in the structure (so were called "hydrohematite" by Wolska, 1981) combined with Fe-vacant sites in the structure. Structural OH increases with decreasing synthesis temperature. It can be recognised by a slight increase in the unit cell size and the intensity ratios of the XRD peaks. This effect becomes evident upon, for example, comparing the intensity (I) of XRD peaks which depend on Fe occupancy (e.g. 012; 104; 110; 024) with that of the 113 peak which depends only on the oxygen atoms. For example, for pure hematite the I_{110}/I_{113} ratio is ca. 4.0 and for $Fe_{1.83}O_{2.5}(OH)_{0.5}$ it dropped to ca. 3.3. Heating removes structural OH/H_2O (Schwertmann et al. 1999). If hematite is prepared from $FeCl_3$ small amounts of Cl may be retained at the surface or within the polydomainic crystals. Kandori et al. (1993) found 0.3 mol% Cl could not be removed by washing; it lowered the point of zero charge from ~pH 8 to ~pH 3.

As mentioned before, the yield of hematite from forced hydrolysis decreases as the degree of hydrolysis increases, i.e. the pH decreases during the formation process. Bao and Koch (1999) showed that the yield increased from ca. 10 to 60% in 0.1–0.2 M $FeCl_3$ solutions as the temperature increased from 35 to 140 °C, and, at a given temperature, it was higher in 0.02 M than in 0.2 M solutions, e.g. 90 vs. 60% at 140 °C.

Because of the similar thermodynamic stability of goethite and hematite, formation of hematite competes with nucleation of goethite which proceeds via dissolution of the ferrihydrite precursor. The lower the temperature, the more likely it is that goethite will form. Goethite nucleation can therefore be inhibited by preheating the oven and also all solutions before they are combined. The presence of chloride should be avoided because it promotes the formation of akaganeite (β-FeOOH). However, chloride concentrations below 0.02 M can be tolerated at temperatures of around 100 °C.

10.4 Monodisperse Hematites of Different Crystal Shapes

Hematites with a particular crystal shape (plates, needles, spindles, pseudocubes, peanuts) and a narrow size distribution (monodisperse) can be obtained by adding various chemicals (shape controllers) to the system. The mechanism behind the control of shape is most likely to be the adsorption of impurities on certain crystal faces thereby reducing their growth rate in favor of that of the other faces. Internally these crystals may be either mono- or polydomainic, depending on the type and concentration of the additive. It must be kept in mind that higher additive concentrations may lead to product contamination. Some examples are summarized in the following section.

Needle-shaped hematite, i.e. crystals with preferential development along the crystallographic *c*-axis, formed after 2–3 d aging of 2-line ferrihydrite (5 mmol Fe in 150 mL solution) at 80 °C in the presence of citrate: 10^{-4} M at pH 10 or 10^{-2} M at pH 12 (Fig. 10-5c) (Schwertmann et al. 1968), or maltose (Cornell, 1985).

The conditions for the formation of pseudocubes, plates, spindles and peanuts, produced mainly by Sugimoto and coworkers are summarized in Tab. 10-1.

Tab. 10-1. Experimental conditions for the production of monodispersed hematites with various crystal shapes

Crystal shape	$FeCl_3$-solution		NaOH		Aging		Reference
	conc. (M)	amount (mL)	conc. (M)	amount (mL)	Temp. (°C)	Duration	
Spheres	0.018	10	0.001§	10	100	24 hr	Matijevic and Scheiner (1978)
Pseudocubes	0.1	100	5.4	100	100	8 days	Sugimoto et al. (1998)
Plates	2.0	40	8	40	180	2 hr	Sugimoto et al. (1996)
Spindles	0.02	1000	0.0003*	1000	100	2 days	Muramatsu et al. (1994)
Peanuts	2.0	100	6.0**	90	100	8 days	Sugimoto et al. (1993)

§ HCl instead of NaOH; * NaH_2PO_4 instead of NaOH; ** plus 10 ml of 0.6 M Na_2SO_4

Fig. 10-6 shows that the pseudocubic and platy hematites produced according to the data given in Table 10-1 are monodispersed. With the ori-

Fig. 10-6. SEM photo of monodispersed platy (upper; Sugimoto et al., 1996; with permission) and pseudocubic (lower; Sugimoto and Sakata, 1992; with permission) hematite produced according to Tab. 10-1 (Courtesy T. Sugimoto).

ginal method, (the condensed Fe^{III} hydroxide gel method), the resulting pseudocubes were polydomainic. Monodomainic crystals were obtained by reducing the Cl concentration to 0.1 M as in Tab. 10-1. The peanut-shaped crystals were polydomainic, whereas spindle-shaped crystals can be either mono- or polydomainic. Polydomainic spindles were produced by aging (1) a ferrihydrite suspension (11.2 g Fe/L) in the presence of 0.02 M oxalate at pH 6 for 165 hr at 70 °C (Fischer and Schwertmann, 1975) (Fig. 10-5 d); (2) an Al-ferrihydrite suspension (Al/Al+Fe = 0.15 mol/mol) at pH 5 and 25 °C for18 yr (Schwertmann et al. 2000) and (3) a mixed $FeCl_3$-Na_2SO_4 solution (Sugimoto et al., 1998); a mixed 0.02 M $FeCl_3$-0.0003 M NaH_2PO_4 solution at 100 °C for 48 hr gave monodomainic spindles (Osaki et al. 1984; Muramatsu et al 1994). Domain character is reflected in the sample surface area: it was about ten times higher for the polydomainic than for the monodomainic sample (140 vs. 13 m^2/g). Other additives investigated were surfactants (Kandori et al., 1995), amines (Kandori et al. 1996), dioxane (Kandori et al., 1998) and dimethylformamide (Kandori et al. 1998 a). Surfactants had essentially no effect on particle shape; cubes and diamond-like crystals were formed in the presence of amides and dioxine and diamond-shaped ones when dimethylformamide was added.

10.5 Other Methods

A number of other methods are listed in the following section. The full details are not provided. Instead, the reader is referred to the appropiate reference.

Decomposition of Metal Chelates. The basic procedure involves aging a solution of Fe^{III} salt in alkaline media (pH 12) in the presence of triethanolamine (TEA) at 250°C for 1 hour (Sapiesko and Matijevic, 1980).

Hydrothermal transformation of various Fe oxides. Ferrihydrite (2-line), lepidocrocite, akaganeite and goethite (if poorly crystalline) can be converted to large (1–3 μm) hexagonal plates of hematite if kept under water in a teflon bomb at 180 °C for 10 days.

Purple hematite. Hematite crystals several μm in size have a purple color. Such crystals can be produced by holding ferrihydrite under very strongly alkaline condition (5 M NaOH) at 70 °C for 8 days. The crystals are idiomorphic. Some goethite is usually associated with the hematite.

10.6 Al-Substituted Hematite

Method

Combine the following volumes (mL) of freshly prepared 0.1 M $Fe(NO_3)_3$ and 0.1 M $Al(NO_3)_3$ solutions in a 500 mL polypropylene bottle:

Fe:	200	198	196	194	190	184	180	176	170	164	160
Al:	0	2	4	6	10	16	20	24	30	36	40

Add sufficient NaOH to each bottle to bring the pH to 7 (complete precipitation), wash the precipitate three times with distilled water, make up to 300 mL total volume with distilled water and adjust again to pH 7 with HNO_3 or NaOH. Store in an oven at 80 °C for 65 days. After crystallization, centrifuge the product, wash with distilled water (or dialyse) and dry.

Product Description

The crystallinity of the Al hematite changes with the level of Al in the structure; a maximum is often obtained at low to medium substitution (5–10 mol%). This corresponds to a minimum in XRD peak broadening and a maximum in γ-ray absorption (Schwertmann et al., 1979; DeGrave et al., 1982; Murad and Schwertmann, 1986). Both X-ray diffraction peaks (Fig. 10-2) and IR absorption bands (Fig. 10-3) show a shift due to Al-for-Fe substitution. The surface area of the hematites is around 40 to 50 m^2/g.

Comments

This hematite has no more than about 1/6 of its Fe replaced by Al. Higher values can be obtained by heating Al goethite to a temperature which is high enough to just dehydrate the goethite to hematite (<500 °C). At even higher temperatures, part of this Al is, however, ejected and forms Al_2O_3.

With low levels of Al in the system (< 5 mol%) the sample contains some goethite whereas at higher Al/(Al+Fe), Al suppresses goethite completely, even at RT (Schwertmann et al. 2000). Goethite can also be avoided by making the system $5 \cdot 10^{-3}$ molar in oxalate at 70 °C. However, the sample will then be contaminated with some oxalate and will need a H_2O_2 treatment to remove it. Another method of lowering the proportion of goethite is to precipitate the mixed solution at 80 °C.

Al-substituted hematites were also produced from Al-containing 2-line ferrihydrite at room temperature and pH 4–7, although at a much lower rate (months to years) (Schwertmann et al. 2000). For the same initial Al/(Al+Fe) ratios the substitution is lower than at higher temperatures. Electron micrographs show rhombic crystals at low substitution (Fig. 10-5 a), identical to unsubstituted ones, and framboidal or spindle-shaped crystals with a grainy or layered interior at higher substitution (Fig. 10-5 b). The latter diffracted X-rays as single crystals.

A 9% Cu-substituted hematite (Fig. 10-5 e) can be obtained by heating a Fe/Cu coprecipitate (Cu/(Cu + Fe) = 0.1) at pH 12° and 90 °C for 60 hr (Cornell and Giovanoli, 1988). The crystals are rhombohedral and the sample surface area ca. 5 m^2/g.

Up to 3 mol% phosphate has been incorporated into hematites; the crystals were spindle-shaped (Galves et al., 1999).

10.7 Coated Hematite

A method of coating monodispersed, spindle-shaped hematite with a uniform layer of SiO_2 is described by Ohmori and Matijevic (1992). Tetraethyl orthosilicate is hydrolyzed in the presence of hematite at 40 °C for between 0.5 and 18 hr to produce SiO_2 layers of different thickness around the hematite crystals.

Hematite can also be coated with yttrium oxide (Aiken and Matijevic, 1987), Al oxide (Kratovil and Matijevic, 1987), Cr hydroxide (Garg and Matijevic, 1988), zirconium oxide (Garg and Matijevic, 1988 a, and Mn oxide (Iul Haq and Matijevic, 1997).

11 Magnetite

11.1 Introduction

Magnetite, $Fe^{II}Fe_2^{III}O_4$ contains Fe^{II} as well as Fe^{III} in its structure. It is, therefore, not thermodynamically stable at atmospheric O_2 pressure. For sufficiently large crystals, such as those of magnetites in rocks, the rate of oxidation by the atmosphere is, however, slow enough so that no protection against oxidation is needed. Synthetic, low temperature magnetite crystals are, on the other hand, often very small (< 100nm) and may, if not protected against atmospheric oxygen, oxidise either during synthesis or thereafter.

There are two basic ways to produce magnetite only the first of which is described here in detail:

(1) by partial oxidation of a Fe^{II} salt solution with KNO_3 under alkaline conditions at 90 °C (David and Welch, 1956; Sidhu et al., 1977), and
(2) by precipitation of a mixed Fe^{II}/Fe^{III} solution with a Fe^{II}/Fe^{III} ratio of 0.5 (Welo and Baudisch, 1925) or of 10 (Taylor et al. 1987).

11.2 Preparation by Oxidation of a Fe^{II} Solution

Method

Dissolve 80 g $FeSO_4 \cdot 7\ H_2O$ in 560 mL deionized water (previously flushed with N_2) in a 1 L glass beaker. The reaction vessel should be closed with an air tight plastic lid containing several holes for inlets and a thermometer (Fig. 11-1). Place the beaker in a water bath (e.g. in a 3 L beaker filled with water at a temperature of 90 °C) and insert the thermometer for temperature control and a gas inlet for purge N_2. The reaction

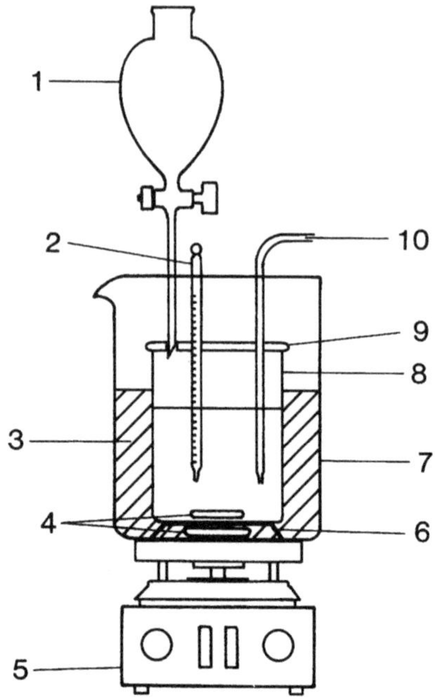

Fig. 11-1. Equipment for magnetite synthesis. 1 Drop funnel, 2 Thermometer, 3 Water bath, 4 Magnet, 5 Stirring and heating instrument, 6 Support for reaction vessel, 7 External beaker used as water bath, 8 Reaction vessel, 9 Plastic lid with inlet holes, 10 Purge gas inlet.

should be carried out under N_2. Once the reaction temperature is reached, add 240 mL of an oxygen-free solution containing 6.46 g KNO_3 and 44.9 g KOH dropwise over approximately 5 min (e.g. through a burette or a drop-funnel). After addition of this solution heat for another 30–60 min, cool overnight and wash the black precipitate. No protection against air oxidation is needed once the reaction is terminated.

Product description

The composition of the resulting magnetite is close to stoichiometric as indicated by chemical analysis ($Fe_{2.08}Fe_{0.92}O_4$), Mössbauer spectroscopy ($Fe_{2.03}Fe_{0.97}O_4$), and unit cell edge length (0.83997(3) nm). The crystals form cubes, bounded by {111} faces and vary in size between 0.05–0.2 µm (Fig. 11-2); the surface area is 4 m^2/g. The XRD lines are sharp

Fig. 11-2. Electron micrograph of magnetite. Bar = 100 nm.

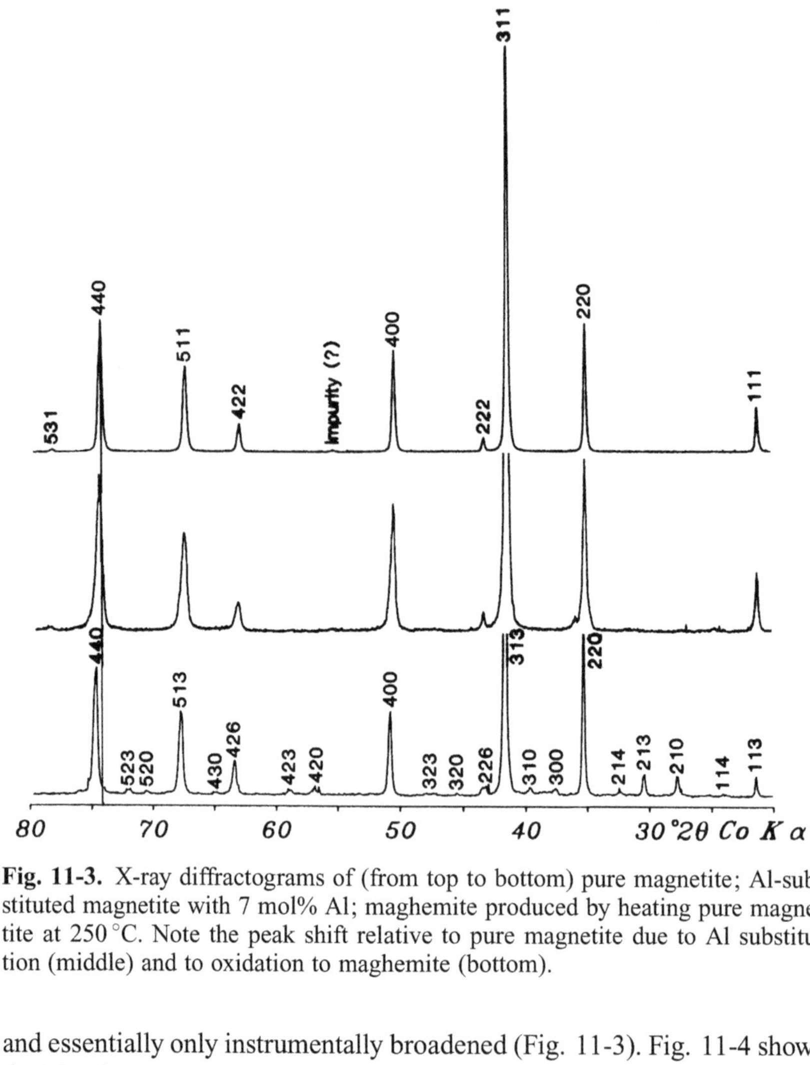

Fig. 11-3. X-ray diffractograms of (from top to bottom) pure magnetite; Al-substituted magnetite with 7 mol% Al; maghemite produced by heating pure magnetite at 250 °C. Note the peak shift relative to pure magnetite due to Al substitution (middle) and to oxidation to maghemite (bottom).

and essentially only instrumentally broadened (Fig. 11-3). Fig. 11-4 shows the Mössbauer spectrum at 298 K.

Comments

$FeCl_2$ (instead of $FeSO_4$) also yields magnetite. Oxidation at RT instead of at 90 °C gave smaller crystals (ca. 50 nm) which chemical analysis, unit cell size and magnetic hyperfine properties showed to be partially

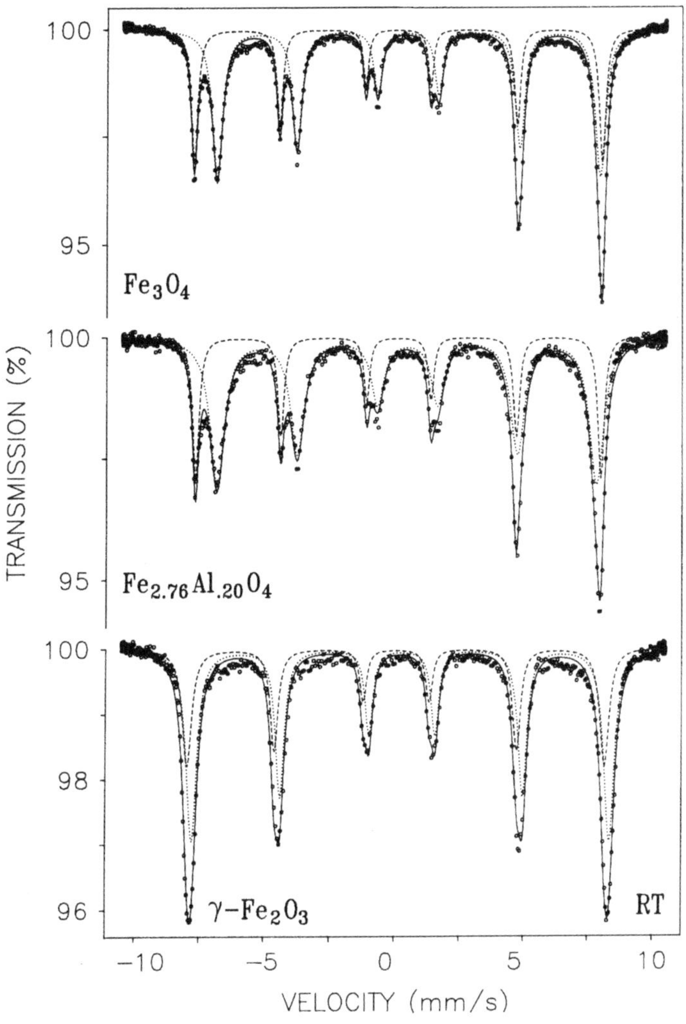

Fig. 11-4. Mössbauer spectra recorded at 298 K of (from top to bottom): magnetite; Al-substituted magnetite with 7 mol% Al; maghemite produced by heating pure magnetite at 250 °C.

oxidized ($Fe^{II}/(Fe^{II}+Fe^{III}) = 0.2-0.3$) immediately after formation (Schwertmann and Murad, 1990); oxidation continued during storage in an air-dried state. After 99 months the above ratio had dropped to 0.146 (Murad and Schwertmann, 1993).

The method of Baudisch and Welo (1925) in which a mixed solution of Fe^{III} and Fe^{II} with a Fe^{II}/Fe^{III} ratio of 2 is reacted at pH 9–10 at RT leads to a poorly crystalline magnetite which is partly oxidized unless air is rigorously excluded. The process probably involves reaction of the Fe^{2+} ions with the 2-line ferrihydrite which precipitates as the pH is brought to pH 9 (Taylor et al. 1987).

Another method of producing magnetite is by reducing hematite in a stream of H_2 at between 250 and 600 °C.

11.3 Cation-Substituted Magnetites

Magnetite with ca. 7 mol% Al substitution can be produced in the same way as described for pure magnetite except that the initial solution contains 10 mol% Al (as $Al_2(SO_4)_3$). The resulting magnetite has the formula $Fe_{0.88}Fe_{1.88}Al_{0.20}\square_{0.004}O_4$ and a reduced cubic unit edge length of 0.8382 nm. The X-ray diffractogram and Mössbauer spectrum are shown in Fig. 11-3 and Fig. 11-4, respectively. The surface area was ca. 20 m^2/g. A range of Al- substituted magnetites synthesized at RT are described in Schwertmann and Murad (1990). The unit cell size a (nm) of these partially oxidized, Al-substituted magnetites (0–14 mol% Al) was related to the level of structural Fe^{II} and Al according to $a = 0.83455 + 0.00693\ Fe^{II} - 0.00789\ Al$. Al-substituted magnetites were also produced at RT by adding NH_4OH (to pH 12) to mixed Fe^{II}-Fe^{III} chloride solutions in which part of the Fe^{III} chloride solution was replaced by $AlCl_3$ (Golden et al. 1994). At Al/(Fe+Al) 0.20 mol/mol, magnetite was the only solid phase whereas above this ratio, goethite and gibbsite also formed.

Magnetites with ca. 10% Ni and 17% Zn in the structure have been produced by Sidhu et al. (1981).

12 Maghemite

12.1 Introduction

Maghemite can be considered a fully oxidized magnetite. In spite of the increase in positive charge upon oxidation of Fe^{II} in magnetite, the cubic structure is preserved. This is achieved through the ejection of 11% of the Fe from the structure thereby creating vacancies (□): $Fe_3O_4 \rightarrow Fe_{2.67}\square_{0.33}O_4$. Where the vacancies are ordered additional XRD peaks occur and the structure can be indexed tetragonally (Fig. 11-3). Upon oxidation, the color changes from black to red-brown. A complete series of transitions exists between magnetite and maghemite.

Maghemite can be produced by heating either lepidocrocite or synthetic magnetite, by oxidation of a mixed Fe^{II}-Fe^{III} solution at RT or by heating ferrihydrite (or other Fe oxides) in the presence of an organic substance.

12.2 Preparation

Method 1
Heat synthetic lepidocrocite (chap. 6) or synthetic magnetite (chap. 11) in a crucible in a furnace for 2 hr at 250 °C.

Method 2
A mixed 0.064 M $FeCl_3$-$FeCl_2$ solution with a Fe^{II}/Fe^{III} of 9 is oxidized in a closed vessel at 20 °C and at a constant pH of 7 using an air flow rate of 10 ml/min (Taylor and Schwertmann, 1974) (For equipment see lepidocrocite synthesis).

Method 3

Mix 1 g of freeze- or air-dried ferrihydrite with 0.3 g ground sugar and heat in a covered crucible at 400 °C for 1 h.

Product Description

The maghemite obtained by heating lepidocrocite is poorly crystalline and has a cubic unit cell edge length of 0.8345 nm; there are no superstructure lines (Schwertmann and Fechter, 1984). The maghemite crystals are acicular like those of the precursor and microporous. The surface area of the product is greater than that of the precursor. Heating of magnetite leads to maghemite with extra X-ray lines (Fig. 11-3), which are sharper the greater the crystallinity of the parent magnetite. Such maghemite has the same morphology and surface area as the magnetite precursor. The Mössbauer spectrum is shown in Fig. 11-4. The maghemites produced by method 2 are usually not pure but contain a few per cent of Fe^{II} ("magnetitic maghemite").

Maghemites derived from ferrihydrite and other OH-containing Fe oxides contain OH and, correspondingly, Fe vacancies in the structure (Stanjek et al. 1999). The maghemite produced by method 3 may contain residual carbon which may be removed by H_2O_2 treatment.

Comments

The heating temperature in method 1 is critical because higher temperatures cause the maghemite to transform to hematite. Crystalline lepidocrocite requires a longer heating time whereas lower temperatures (180–200 °C) are sufficient for poorly crystalline, synthetic lepidocrocites. The product obtained by method 2 is very sensitive to experimental conditions. Increasing the rate of oxidation leads to the formation of lepidocrocite and higher Fe^{III}/Fe^{II} ratios produce ferrihydrite. In contrast, higher temperatures (e.g. 60 °C) favor maghemite (Taylor and Schwertmann, 1974).

Monodispersed spindle-type maghemite can be produced by first transforming spindle-type hematite (see chap. 10) into magnetite (by heating the hematite in a H_2 gas stream at 340–400 °C for 1–3 h) and then oxidizing the magnetite to maghemite with air at 240 °C for 1–2 h (Ozaki and Matijevic, 1985).

13 Green Rusts

13.1 Introduction

Green Rusts are $Fe^{II,III}$ hydroxy salts with a layer structure in which hexagonally closed-packed, brucite-type $Fe(OH)_2$ layers carry a positive charge due to some Fe^{III}. The charge is balanced by interlayer anions, such as chloride, sulfate and carbonate, but other anions (Br^-, I^-, ClO_4^-, NO_3^-) may also be intercalated (Lewis, 1997). Green rusts are synthesized by slow oxidation of a Fe^{II} solution at a pH of around 7 (Taylor et al., 1985; Schwertmann and Fechter, 1994; Lewis, 1996), by interacting Fe^{2+} ions with ferrihydrite under near-neutral conditions (Bruun-Hansen et al. 1994) or by oxidation of freshly precipitated $Fe(OH)_2$ at a pH at which the solution is supersaturated with respect to $Fe(OH)_2$ (Feitknecht and Keller, 1950; Genin et al. 1998). Green rust often appears as an intermediate product during the formation of FeOOH from Fe^{II} systems.

Green Rusts are very sensitive to oxidation which complicates their synthesis and stabilization. Thermodynamic data for sulfate green rust are given in Bruun Hansen et al. (1994) and E_h-pH diagrams as well as thermodynamic data for chloride, sulfate and carbonate green rusts in Genin et al. (1998). Green rusts in which part of the Fe^{III} is substituted by Al were produced by Taylor and McKenzie (1980) and one in which part of the Fe^{II} is replaced by Ni^{II}, by Refait and Genin (1997).

13.2 Preparation

Place 370 mL of 0.1 M $FeCl_2$ or $FeSO_4$ solution, freshly prepared from $FeCl_2 \cdot 4\ H_2O$ or $FeSO_4 \cdot 7\ H_2O$, and purged with N_2 (or Ar; Bruun-Hansen et al., 1994) in a closed reaction vessel equipped with a magnetic stirrer, a gas and base inlet and a pH electrode connected to an automatic titrator and a pH meter. Titrate to pH 7.0 (higher pH values may lead to magnetite formation). Then replace the nitrogen gas by CO_2-free air at a rate of ca. 7ml/min and keep the pH constant by adding 1 M NaOH. After ca. 40 min a dense green precipitate forms; this must be collected by freeze-drying under a O_2 free atmosphere.

Lewis (1996) suggests the following freeze-drying procedure: transfer the product rapidly into a polyethylene centrifuge tube, flush with nitrogen, close with a cap and centrifuge for 1–2 min at 1000 rpm. Decant the clear liquid, add a small volume of liquid nitrogen to freeze the sediment and seal with a rubber stopper fitted with a stopcock ready for attachment to a port of a freeze drier. After freeze-drying, the dried powder could be kept sealed in the tube under vacuum. For X-ray diffraction and Mössbauer spectroscopy, moist samples, i.e. straight after centrifugation, may also be used, if a vacuum cell at the X-ray instrument is available.

Fig. 13-1 shows a Mössbauer spectrum of the carbonate form of green rust with two separate doublets fitted to both the Fe^{II} and Fe^{III} doublet.

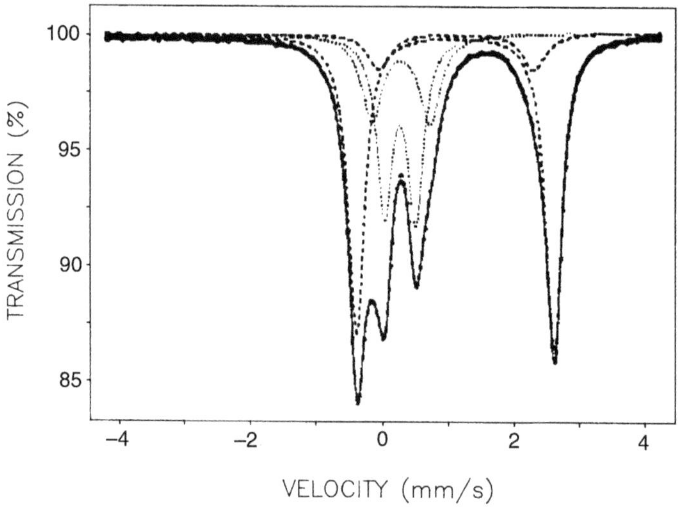

Fig. 13-1. Mössbauer spectrum of carbonate green rust recorded at 120 K and fitted to two Fe^{III} and two Fe^{II} doublets. Courtesy E. Murad (see Murad and Taylor, 1984).

Product description

The green rusts form large platy hexagonal crystals. An X-ray diffractogram of sulfate green rust is shown in Fig. 13-2; the very sharp lines show the product is highly crystalline. Figure 13-2 also documents the gradual oxidation of sulfate green rust at pH 7.5 over ca. 4 hr to lepidocrocite and some goethite.

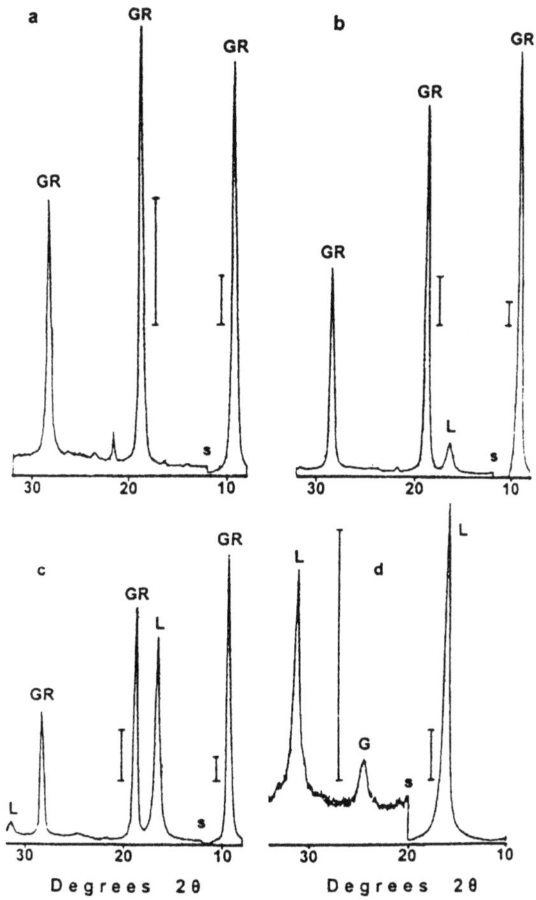

Fig. 13-2. X-ray diffractograms (CoKα-radiation) of sulfate green rust (GR) and the stages in its oxidation by air to lepidocrocite (L) and goethite (G). a: 0%; b: 15%; c: 75% and d: 100% oxidation. Points marked 's' indicate a change in scale for X-ray intensity. Bar represents 1000 cps (Lewis, 1997; with permission).

14 Schwertmannite

14.1 Introduction

Surface waters draining pyritic rocks, mines and mine spoils are usually very acid (pH <3) and rich in dissolved iron and sulfate. As the pH increases to 3–4 due to admixture with uncontaminated water or to buffering with soil and rock material, an ochreous precipitate is often formed following oxidation of Fe^{2+} by the bacterium *Thiobacillus ferrooxidans*. A compound hitherto unknown was identified recently in these precipitates (Bigham et al. 1990). It is a Fe^{III} oxyhydroxy sulfate with a somewhat variable composition and an ideal formula of $Fe_8O_8(OH)_x(SO_4)_y$, where $8-x = 2y$ and $1.0 < y < 1.75$. It has been recognized by the International Mineralogical Association as the mineral schwertmannite (Bigham et al. 1994). The structure is similar to that of akaganéite. Schwertmannite differs from akaganéite in its poor crystal development in the crystallographic [110] direction and its X-ray pattern, therefore, lacks the strongest (110) line of akaganéite. Furthermore, sulfate instead of chloride is incorporated in the tunnels where the anion is believed to share some oxygen atoms with the Fe(O,OH) octahedra.

This compound can be synthesized in a simple inorganic manner from a sulfate- containing Fe^{III} solution (see below). If an acid $FeSO_4$ solution is used, oxidation can only be carried out in a controlled bioreactor using the bacterium *Thiobacillus ferooxidans* (Plate VIII; Bigham et al., 1990).

14.2 Preparation

Method

Preheat 2 L of distilled water to 60 °C in an oven, quickly add 10.8 g $FeCl_3 \cdot 6\,H_2O$ and 3 g of Na_2SO_4 (1000 mg SO_4/L) and heat for a further 12 min at 60 °C. After cooling at room temperature, dialyse the suspension for a period of 30 days and finally freeze-dry the solid.

Product description

The method produces ca. 2 g of a light ochreous powder which has an XRD pattern with ca. 8 broadened lines (Fig. 14-1) and a surface area of ca. 250 m^2/g. The SEM photo shows hedge-hog type crystals (Fig. 14-1) and TEM indicates that the spicules are only a few nm across (Fig. 14-2). The IR-spectrum shows intense bands of SO_4^{2-}. The Mössbauer spectra taken at various temperatures (Fig. 14-3) show the gradual development of a sextet, i. e. magnetic ordering at between 60 and 4.2 K.

Comments

Schwertmannite is stable with respect to ferrihydrite in acid, sulfate-rich waters but metastable with respect to goethite. In water, schwertmannite transforms to goethite by a hydrolytic dissolution reaction:

$$Fe_8O_8(OH)_6(SO_4) + 2\,H_2O \longrightarrow 8\,FeOOH + H_2SO_4$$

As seen from this equation the pH drops and counteracts the transformation. If the pH is raised towards neutral, as in neutralization ponds, ferrihydrite will form. In an air-dry state schwertmannite can be stored unchanged for years.

A complete series between schwertmannite in the pure sulfate system and akaganeite in the pure chloride system has been synthesized from Fe^{III}-solutions with various Cl^-/SO_4^{2-} ratios (Bigham et al. 1990). An ion activity product (IAP) for the solubility of schwertmannite, log IAP = 18.0 ± 2.5, was calculated from the chemical composition of a number of mine drainage solutions by Bigham et al. (1996).

An analogous compound with an akaganéite structure but with nitrate in the tunnels was prepared as follows (Schwertmann et al. 1996): Add 20.2 g of $Fe(NO_3)_3 \cdot 9\,H_2O$ (0.025 M) to 2 L of water preheated to 75 °C

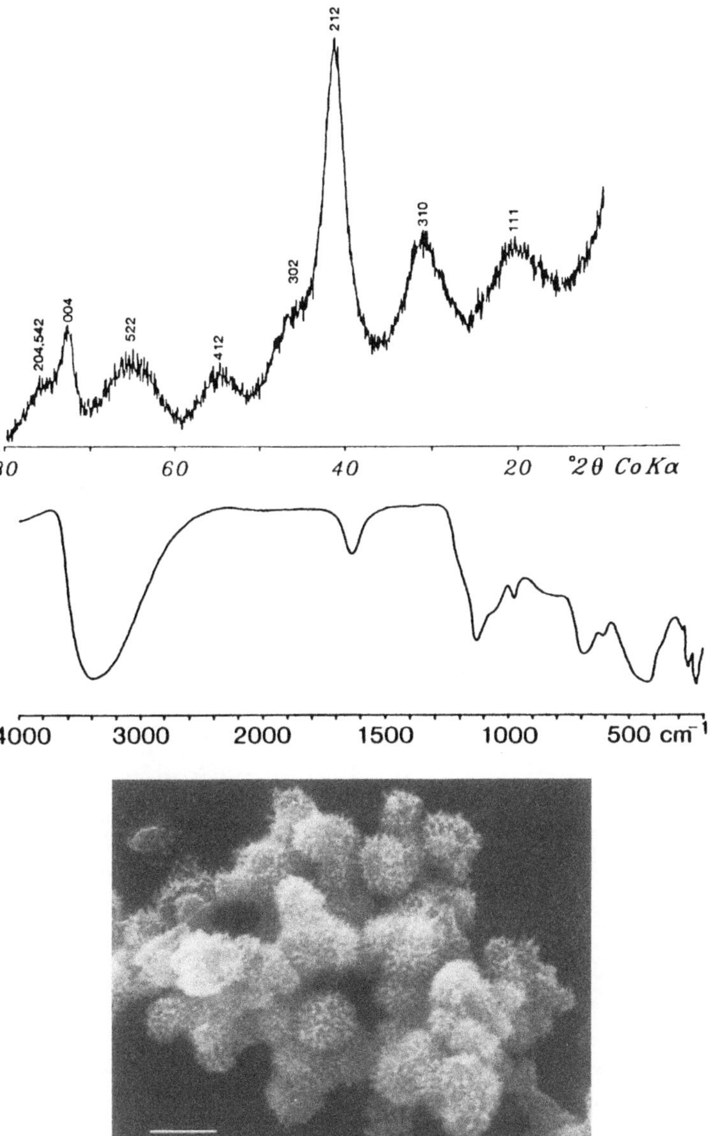

Fig. 14-1. X-ray diffractogram, IR spectrum, and a scanning electron micrograph of schwertmannite (Bigham et al., 1990; with permission). Bar = 1 µm. (TEM – Courtesy J. M. Bigham)

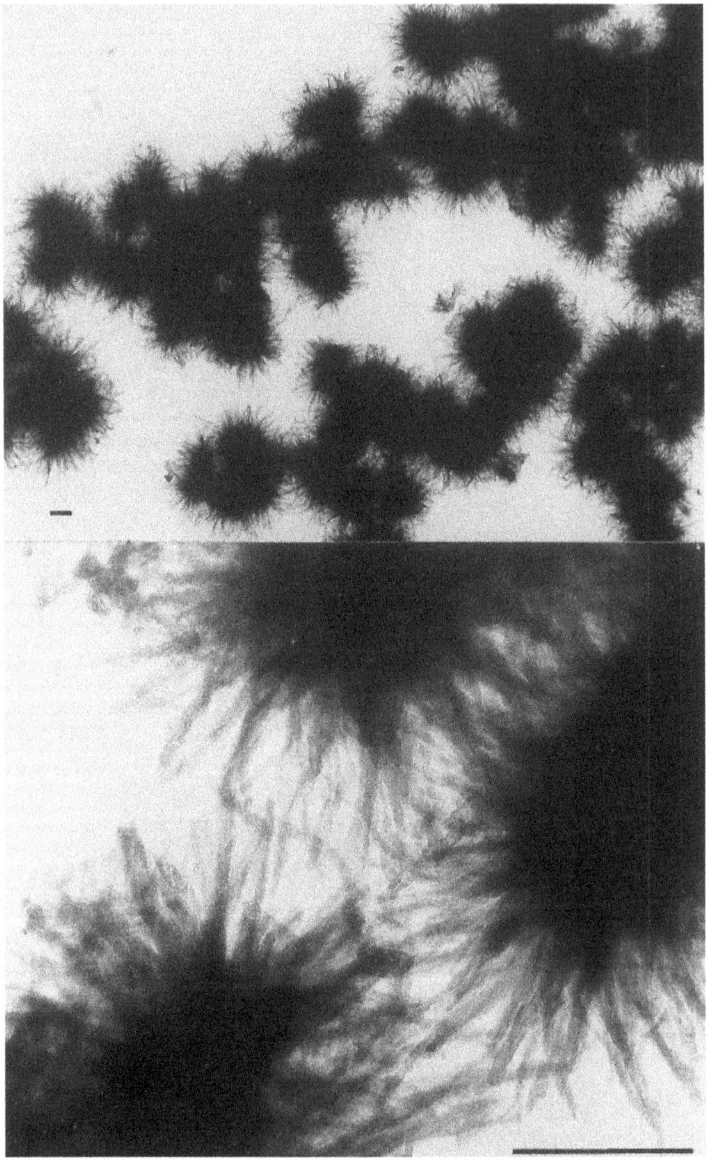

Fig. 14-2. Transmission electron micrograph of schwertmannite (upper: Bigham et al. 1990; with permission). Bar = 100 nm. (Lower: Courtesy H.-Ch. Bartscherer, Munich)

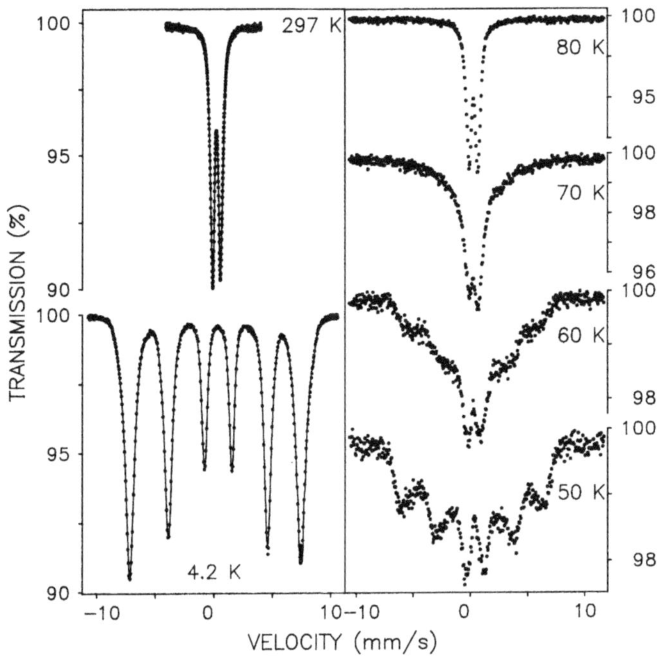

Fig. 14-3. Mössbauer spectra of schwertmannite recorded at 297 K (upper left); 4.2 K (lower left) and in the temperature range between 80 and 50 K (right) to demonstrate the gradual magnetic ordering that occurs with decreasing temperature (Bigham et al. 1990; with permission).

and keep at this temperature for 12 min. Shock-freeze the red sol and freeze-dry. The dry product shows five, broad X-ray peaks between 0.255 and 0.147 nm which correspond to the peaks of schwertmannite. The nitrate content varies between 17.5 and 18.6 weight% corresponding to a tentative formula of $FeO(OH)_{1-x}(NO_3)_x$ with $0.2 < x < 0.3$. In water the compound hydrolyses to 6-line ferrihydrite and the nitrate is released.

Further candidates for the tunnel position in the schwertmannite structure is selenate. It seems that the oxyanions of hexavalent elements (S, Se, Cr) can be accomodated in the akaganeite structure whereas those of pentavalent ones (P, As) may only adsorb on the surface (Waychunas et al. 1995).

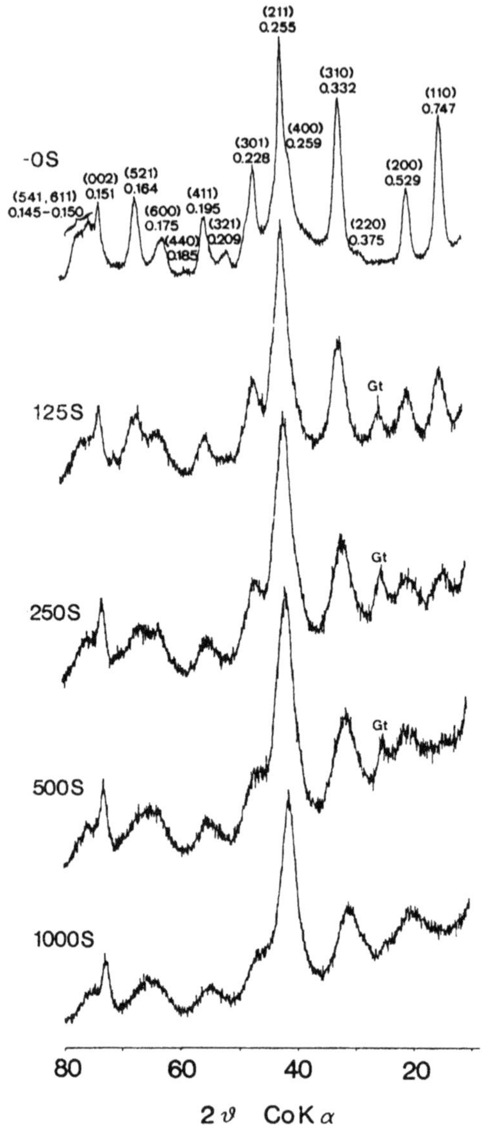

Fig. 14-4. X-ray diffractograms from products prepared by hydrolysis of 0.02 M $FeCl_3$ in the presence of 0, 125, 250, 500, and 1000 µg/ml SO_4^{2-}. A continuous series between akaganeite at zero SO_4^{2-} and schwertmannite at 1000 µg/ml SO_4^{2-} is formed. Gt indicates a trace of goethite (Bigham et al. 1990; with permission).

15 Coating of SiO_2 Sand (Quartz; Cristobalite) with Iron Oxides

15.1 Introduction

Coating of SiO_2 sand with well defined iron oxides involves a heterogeneous reaction in which the SiO_2 sand is mixed with an aqueous suspension of the iron oxide. The coating reaction can be carried out in the laboratory with different iron oxides (goethite, hematite, lepidocrocite, ferrihydrite and maghemite) using different SiO_2 phases (cristobalite and quartz).

In view of the frequent ocurrences of iron oxide coated mineral and rock particles in nature (Schwertmann and Friedl, 1998), sand coated with well defined iron oxides provides a useful model system for studying adsorption and dissolution reactions with the iron oxide in a mechanically stable form. In such a system the overall reactivity of the coated material is strongly dominated by the reactivity of the iron oxides. One interesting application of this material is as a model column matrix for transport studies. For example, cristobalite sand coated with goethite as described below, has been used as a column matrix to study reactive transport behaviour of protons (Scheidegger et al., 1994; Bürgisser et al., 1994), fluoride (Meeussen et al., 1996), sulfate (Meeussen et al., 1999), phosphate (Geelhoed et al., 1997), and uranyl ions (Gabriel et al., 1998). Ferrihydrite coated quartz sand has been used in a similar study for investigation of the redox dynamics of organic pollutants in a column subjected to a fluctuating water table (Sinke et al., 1998). Ferrihydrite coated cristobalite sand has been used in a study of the bio-transformation of organic compounds attached to cells in a percolation column operated at various flow rates and biomass contents (Tros et al., 1998). Finally, the coating procedure has been applied to mount iron oxide particles firmly on quartz for atomic force microscopy (AFM) examination (Weidler et al., 1996).

15.2 Preparation

Method (Scheidegger et al. 1993)
Mix 100 mg of iron oxide with 10 ml $NaNO_3$ solution (ionic strength: 0.001–1 M) in a 50 ml polyethylene tube and adjust to the desired pH (see below) with dilute HNO_3. To obtain a homogeneous suspension shake the mixture at 25 °C for 24 h. Then add 2.5 g of sand and shake the mixture for another 24 h. When the sand has settled, measure the pH and resuspend the supernatant several times in a $NaNO_3$ solution with an ionic strength and pH equal to that of the reaction medium. Finally, wash the coated sand free from any unattached iron oxide first with the salt solution and then with pure water using a nylon sieve (63 μm). Owing to the poor dispersibility of the iron oxides near the point of zero charge (PZC), samples with a pH of between pH 5 and 9 should, in addition, be washed with 1 M $NaNO_3$ solution of pH 3 (adjusted with HNO_3). This procedure allows the coated sand with strongly bound iron oxide particles to be separated from weakly attached iron oxide aggregates. Dry the coated sand in air or in an oven at 110 °C (not for ferrihydrite) for 24 h. The amount of bound goethite can be determined by dissolving the sample completely in a strong acid and measuring the [Fe].

Product Description
Figure 15-1 shows scanning electron microscope images of goethite- and hematite-coated cristobalite sand produced following the above method using a pH of 2.5. The goethite coated sample contains 14.4 mg/g sand and the surface area increased from 0.08 to 0.38 m^2/g. Experiments with cristobalite sand showed that once the sand is coated with goethite, the iron oxide particles adhere very strongly to the SiO_2 surface. Neither common dispersing agents (such as calgon) nor strong acids and bases (1 M HNO_3, 10 M NaOH) could detach the particles. Of course, any agent able to dissolve chemically either the pure goethite or the pure sand will also attack the coated sand (e.g. concentrated HCl, or an aqueous solution of $Na_2S_2O_4$). Ultrasonic treatment of coated sand was the only procedure which led to partial detachment of the goethite particles.

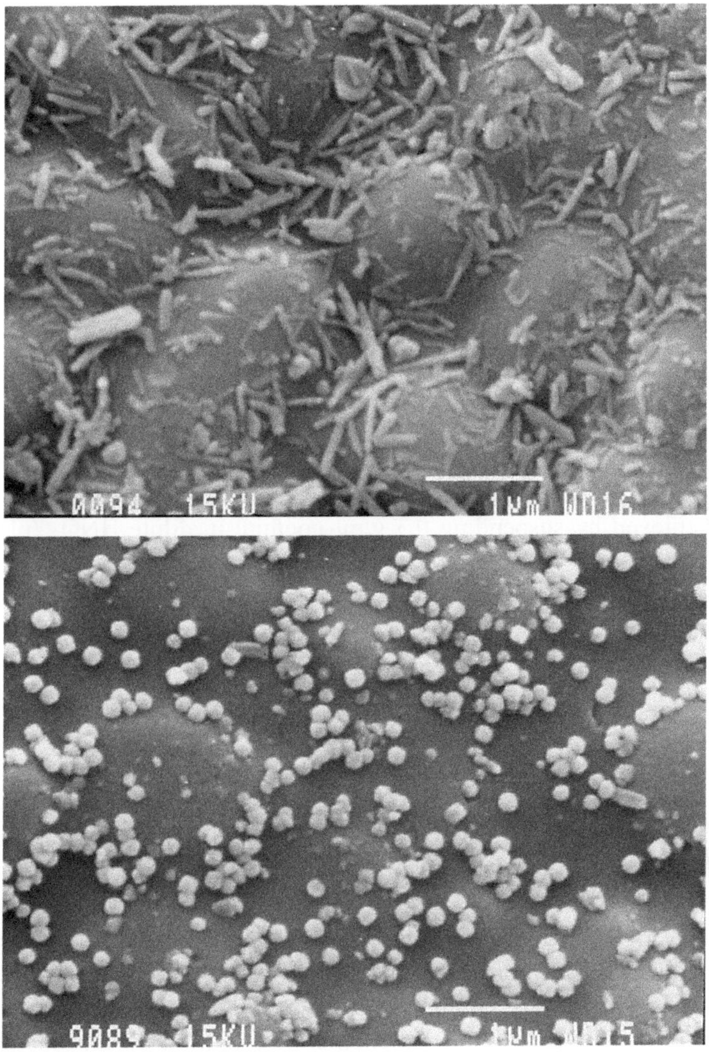

Fig. 15-1. Scanning electron micrographs of goethite- (upper) and hematite-coated (lower) cristobalite produced at pH 2.5. The amounts of iron oxide attached are 14.4 (goethite) and ca. 1 mg (hematite) per g cristobalite sand. (Courtesy A. Scheidegger). (From Scheidegger et al., 1993; with permission.)

Comments

Detailed investigations into the coating of cristobalite with goethite demonstrated that the extent of the coating increases with increasing pH up to the PZC of goethite (ca. pH 8) and decreases abruptly at pH values above the PZC (Scheidegger et al., 1993). This behavior can be explained using a simple electrostatic model, based on the assumption that the charge of the adsorbed goethite particles neutralizes the charge of the SiO_2 surface. Although the coating is maximal at the PZC, the distribution of the Fe oxide crystals over the surface is less homogenous at this pH. Lower pH values are, therefore, recommended. The extent of coating also shows a marked dependence on ionic strength, iron oxide to sand ratio and temperature. A maximum was observed near an ionic strength of 0.01 M. Higher goethite to cristobalite ratios in the reaction mixture and an increase in temperature resulted in higher coating densities. The highest coating density (14.4. mg goethite/g cristobalite sand) was achieved by suspending 0.3 g/ml goethite in dilute HNO_3 (pH 2.5). The very dense suspension was shaken for 24 h and 1.8 g cristobalite sand/ml suspension was added. The mixture was then enclosed in a high pressure vessel and put into an oven at 120 °C for 24 h, frequently taken out of the oven and shaken, slowly cooled down and washed.

Coatings with 2-line ferrihydrite (approx. 6 mg/g cristobalite), hematite (approx. 1 mg/g cristobalite), lepidocrocite (approx. 0.5 mg/g cristobalite) and maghemite (approx. 3 mg/g cristobalite) were obtained at pH 2.5. For these oxides the pH-dependence of the coating reaction seems comparable to that for goethite on cristobalite. Goethite coated quartz sand has been obtained at pH 7 with a coating density of approx. 3.5 mg goethite per g quartz sand. This coating density is comparable to that obtained for the goethite cristobalite system under the same reaction conditions. The coating of a synthetic, porous siran carrier material (sodium orthosilicate) with goethite (approx. 4 mg/g siran carrier, reaction medium pH 2.5) shows that the coating phenomenon is not limited to cristobalite and quartz sand alone.

16 An Experimental Lecture for Students on the Formation of Iron Oxides

16.1 Introduction

Eight different iron oxides have been identified in the weathering environment to date. They are: goethite and hematite (common), lepidocrocite, maghemite, ferrihydrite and magnetite (moderately widespread) and akaganeite and feroxyhyte (rare). These minerals often occur in close association. Since they have similar or even identical composition, a logical question is – under what conditions do the different oxides form?

The pathways by which the different Fe oxides form are complex and not yet fully understood. Hence, the teaching of how Fe oxides form can be a challenge. The use of visual aids can help illustrate and clarify the topic and fortunately, the Fe oxides lend themselves well to this approach. In the first place, the different forms can be recognized and identified by their striking colors (yellow, brown, red and black) and secondly, as shown in previous chapters, they are easily and rapidly synthesized in the laboratory.

By varying the conditions of synthesis much can be learned about the experimental parameters which govern iron oxide formation during weathering and soil formation (pedogenesis) (Schwertmann, 1985; 1988a; Fitzpatrick, 1988; Schwertmann and Taylor, 1989).

This chapter consists of a 45 minute lecture/demonstration of iron oxide formation in vitro and in situ. Several simple bench syntheses of iron oxides are carried out and explained. The lecture may be concluded with a short video which illustrates iron oxide occurrence in a real landscape.

16.2 Demonstration: Synthesis of Fe Oxides

In this section, five important Fe oxides, namely goethite, hematite, lepidocrocite, magnetite and ferrihydrite are produced simultaneously within 25–30 min using simple apparatus, namely: 1 L beakers, pH electrode, pH meter, gas inlet, air or oxygen supply, dropping funnel, gas washing bottle, hot plate with magnetic stirrer and various chemicals.

Throughout each experiment, the solution/suspension must be stirred constantly, preferably using an overhead stirrer. The experimental equipment must be installed beforehand and the solution for hematite synthesis already heated to close to 90 °C. The syntheses must be started immediately and while the oxides form, the pathways of formation are discussed.

The first three oxides are prepared by oxidation of Fe^{II} solutions under slightly different conditions of pH, Fe concentration and rate of oxidation. All preparations are carried out in a 1 L beaker.

Goethite: Dissolve 3 g $FeCl_2 \cdot 4\,H_2O$ in 300 mL distilled water, then add 33 mL $NaHCO_3$ solution. A grayish-green precipitate forms. After ca. 25 minutes, air oxidation is complete and the precipitate turns yellowish.

Magnetite: Dissolve 12 g $FeCl_2 \cdot 4\,H_2O$ with vigorous stirring in 300 mL distilled water, then add 115 mL M NaOH in one lot. A blueish precipitate forms and after 25 min atmospheric oxidation, this material turns black and sticks to a magnet.

Lepidocrocite: Dissolve 12 g $FeCl_2 \cdot 4H_2O$ with vigorous stirring in 300 mL distilled water. The beaker should be equipped with a glass electrode connected to a pH meter, a gas inlet connecting an air or oxygen cylinder and a dropping funnel (outlet above the solution level) containing 125 mL M NaOH. Adjust the pH of the system to 6.5–6.8 by adding NaOH dropwise, then open the gas cylinder and aerate the suspension (ca. 0.1–2 L/min; the rate may be controlled by using a gas washing bottle). The initial greenish black precipitate becomes orange after 20 min. Throughout the reaction, the pH of the suspension must be maintained at 6.5–6.8 by adding NaOH from the dropping funnel as needed.

Hematite: Bring 300 mL distilled water in a 1 L beaker to 90 °C. The beaker stands on a magnetic hot plate. When the required temperature is reached, add 31 g $Fe(ClO_4)_3 \cdot 9\ H_2O$ and maintain the temperature while stirring. The solution gradually becomes wine red and after 30 min. a red precipitate is visible.

Hematite can also be prepared from 2-line ferrihydrite. This demonstration should be started the previous day. The suspension of ferrihydrite must be refluxed at 90–100 °C for 30 hr after which there is a color change from the dark reddish brown of the ferrihydrite to the deep blood red of the hematite.

Ferrihydrite (2-line): Dissolve 40 g $Fe(NO_3)_3 \cdot 9\ H_2O$ (or the appropriate amount of any other Fe^{III} salt) in 500 mL distilled water, then add, with stirring, ca. 330 mL M KOH to bring the pH to 7–8. Ferrihydrite immediately forms as a voluminous, dark, reddish brown precipitate which settles rapidly.

The more crystalline, 6-line ferrihydrite cannot be synthesized during the lecture because it needs a long period for dialysis (see Chap. 8.2).

16.3 Lecture: Processes by which Fe oxides form

Once the syntheses are underway, the lecture can be presented. The graphs for this section as well as the list of oxides in the introduction should be available for presentation as slides or overheads. It is worth using colors for the different Fe oxides and these should be mnemonic and of course, the same on all graphs, namely, black for magnetite, red for hematite, brown for ferrihydrite, yellow for goethite and orange for lepidocrocite (Schwertmann, 1993).

The primary process by which Fe oxides form in nature involves oxidative, hydrolytic release of Fe^{II} mostly from primary Fe^{II} silicates (Fig. 16-1). In the pH range of a normal weathering environment and under aerobic conditions, the Fe^{III} oxides are very stable and persist over long time spans. In an anaerobic environment, however, Fe^{III} oxides may be reductively dissolved by micro-organisms through enzymatic transfer

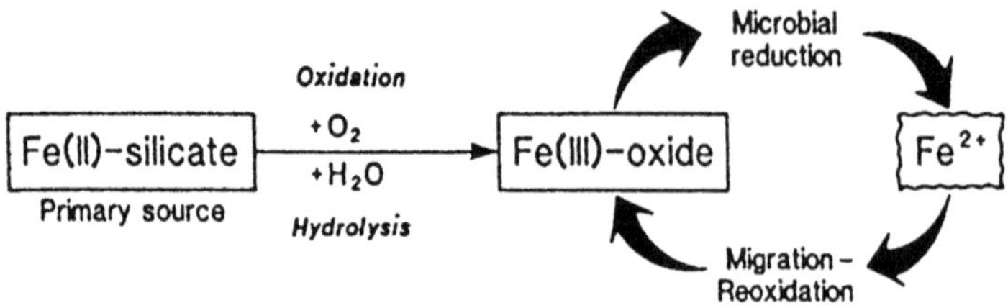

Fig. 16-1. Basic processes of Fe oxide formation in nature. (Schwertmann et al., 1995; with permission)

of electrons from the decomposing biomass (this process is termed iron respiration);

$$4\ Fe^{III}OOH + \underset{\text{biomass}}{CH_2O} + 7\ CO_2 + H_2O \longrightarrow 4\ Fe^{II}(HCO_3)_2$$

Because Fe is mobilized by reduction, it can be redistributed over a range of distances within a soil horizon, a soil profile or even a soilscape. This characteristic together with the fact that there are several different Fe oxides, results in numerous occurrences of these compounds in weathering environments. This can be illustrated with color slides, if time is available.

Although the type of mineral that ultimately forms is mainly governed by its thermodynamic stability, what is observed at any time or even as a quasi-stable end product, is governed by kinetic factors. Important parameters that govern the phase that forms are: temperature; solution composition (pH; Eh; Fe concentration and oxidation state; type and concentration of anions); rate of Fe supply; rate of oxidation.

Thermodynamic stability which varies for the different Fe oxides is the most important parameter if the system has reached equilibrium. The two most stable oxides are goethite and hematite and indeed, these are, by far, the oxides most frequently found in the weathering environment. Thermodynamic data tell us, however, that the formation of goethite or hematite involves almost identical reaction energies, so this data is not sufficient to decide unequivocally which compound is the more stable. Thus kinetic factors come into play. These are far from being fully understood, although the general conditions that favor certain Fe oxides have been established. For example, higher temperature and/or lower water activity favor formation of hematite over that of goethite because these conditions promote dehydration. Nevertheless, hematite can easily form at a water activity of 1 (liquid water). In fact, the transformation of 2-line ferrihydrite into hematite needs a minimum of water in the system (Schwertmann et al. 1999). Other factors which determine which phase forms are pH and solution components such as Al, Si and organic compounds.

The occurrence and persistence of mestable oxides such as lepidocrocite and ferrihydrite can also be explained by kinetic factors. Some of these play a significant role in the natural environment. The synthesis experiments with temperatures of 25 and 90 °C, pH values of 6 and 8, various anions (chloride, chlorate, sulfate) and two different initial oxidation states of Fe illustrate this.

Since pH is an important factor, the protons that are produced during the synthesis process by hydrolysis of Fe^{3+}, must be taken into account. The reactions involved are:

Step 1. Oxidation:	$4\,Fe^{2+} + O_2 + 4\,H^+$	$\longrightarrow$	$4\,Fe^{3+} + 2\,H_2O$
Step 2. Hydrolysis:	$4\,Fe^{3+} + 8\,H_2O$	$\longrightarrow$	$4\,FeOOH + 12\,H^+$
Net reaction:	$4\,Fe^{2+} + O_2 + 6\,H_2O$	$\longrightarrow$	$4\,FeOOH + 8\,H^+$

In the weathering environment, the protons may be buffered by solid-phase buffer systems. In our experiments, the solutions are buffered by $NaHCO_3$ (goethite system), by adding an initial excess of base (magnetite system) or by using a pH-stat device (lepidocrocite system).

Hematite and ferrihydrite are formed via hydrolysis of Fe^{III} solutions. To produce hematite rapidly in an aqueous system, an $Fe(ClO_4)_3$ solution is hydrolyzed at 90 °C and low pH (ca. 2). With more time available (months–years) hematite may also form in aqueous solution at ambient temperature and in fact, even at temperatures as low as 4 °C.

The very poorly ordered ferrihydrite with only two, broad XRD peaks is the result of extremely rapid hydrolysis of Fe^{3+} ions at pH >5. The better ordered ferrihydrite with 6–7 XRD peaks can be obtained by a method described in section 8.4.

Figure 16-2 summarizes the pathways and conditions for the formation of the five oxides in the above experiments.

The different oxides can be distinguished by their colors. The purity can be checked using X-ray diffraction analysis. Diffractograms of the var-

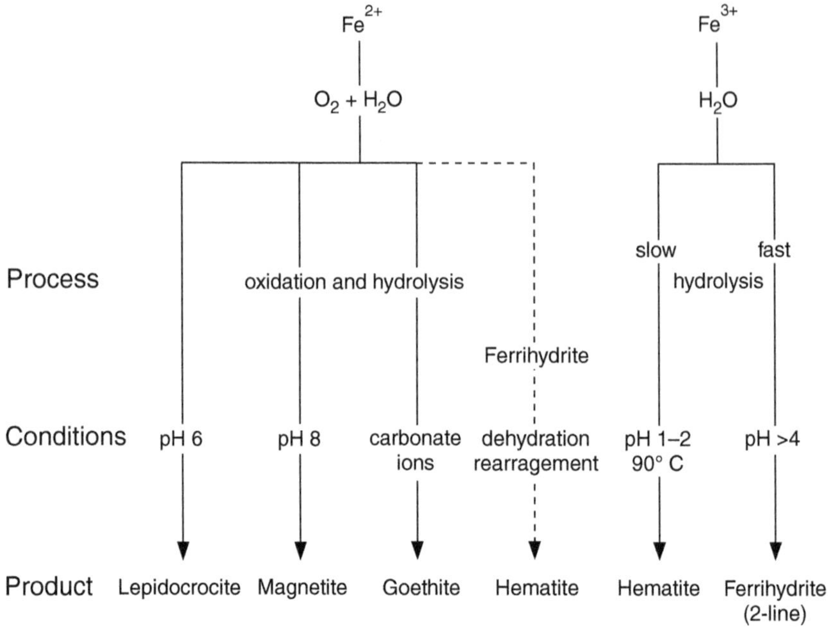

Fig. 16-2. Formation pathways and conditions for synthesis of Fe oxides in the experiments (full lines; broken line indicates pathway not shown in the lecture). (From Schwertmann et al., 1995; with permission.)

ious Fe oxides are shown throughout the book. As seen from the sharpness of the X-ray peaks, the magnetite and the hematite are extremely crystalline (sharp peaks), whereas the goethite is poorly crystalline (broad peaks) and the lepidocrocite is intermediate.

16.4 Video: Iron in a Landscape

The lecture can end "in the field" with a five min. video (obtainable from Dr. M. Stross, Medienzentrum, Techn. Universität, Lotstr.; D-80290 München; Germany). This video concentrates on a low moor soil profile in a Pleistocene valley near Munich in which two iron-rich layers –

a layer of whitish siderite ($FeCO_3$) capped by a reddish brown goethite layer – occur. The siderite is identified by a color test for Fe^{2+} involving reaction with 2,2 bipyridyl – a red color develops (Childs, 1981). The goethite cap has formed by oxidation of the siderite with atmospheric oxygen. Also illustrated is the behavior of the goethite layer upon burning the neighboring peat (this often occurred in past centuries for reclamation purposes). It is seen that upon firing, the yellow goethite turns into bright red hematite. Due to the presence of the organic matter (peat), maghemite forms as well as hematite. The presence of the maghemite can be easily shown because it sticks to a magnet. This experiment is also suitable for a benchtop demonstration.

Further details and explanations are provided in the video itself and in McMillan and Schwertmann (1998).

This chapter is based on a lecture given at the 10th International Clay Conference, Adelaide, Australia, in 1993 and published in the Proceedings of this Conference (1995, G. J. Churchman et al., Eds.) Clays Controlling The Environment p. 11–14. With permission of CSIRO Publishing, Melbourne.

References

Aiken, B., Hsu, W. P. and Matijevic, E. Preparation and properties of coated uniform colloidal particles of lanthanide compounds. III. Y(III) and mixed Y(III)/Ce(III) systems. J. Amer. Ceram. Soc. 71: 845–853.

Allen, T. (1990) Particle size measurement. Chap. 4; 4th Ed. Chapman and Hall, London.

Atkinson, R. J., Posner, A. M. and Quirk, J.P. (1968) Crystal nucleation in Fe(III) solutions and hydroxide gels. J. Inorg. Nucl. Chem. 30: 2371–2381.

Atkinson, R. J., Posner, A. M. and Quirk, J. P. (1977) Crystal nucleation and growth in hydrolysing iron(III) chloride solutions. Clays Clay Min. 25: 49–56.

Bailey, J. K:, Brinker C. J. and McCarthey, M. L. (1993) Growth mechanism of iron oxide particles of different morphologies from forced hydrolysis of ferric chloride solutions. J. Coll. Interface Sci. 157: 1–13.

Bao, H. and Koch, P. L. (1999) Oxygen isotope fractionation in ferric oxide-water-systems: Low temperature synthesis. Geochim. Cosmochim. Acta 63: 599–613.

Barron, V. and Torrent, J. (1984) Influence of aluminum substitution on the color of synthetic hematites. Clays Clay Min. 32: 157–158.

Battle P. B. and Cheetham, A. K., (1979) The magnetic structure of non-stoichiometric ferrous oxide J. Phys.C Solid State Phys.12: 337–345.

Baudisch, O. and Albrecht, W. H. (1932) Gamma ferric oxide hydrate. J. Amer. Chem. Soc. 54: 943–947.

Bender Koch, C., Oxborrow, C. A., Morup, S., Madsen, M. B., Quinn, A. J., and Coey, J. M. D. (1995) Magnetic properties of feroxyhyte (δ'-FeOOH). Phys. Chem. Min. 22: 333–341.

Bernal, J. D., Dasgupta, D. R. and Mackay, A. L. (1959) The oxides and hydroxides of iron and their structural interchanges. Clay Min. Bull. 4: 15–30.

Bigham, J. M., Carlson, L. and Murad, E. (1994) Schwertmannite. a new iron oxyhydroxysulfate from Pyhäsalmi, Finland, and other localities. Min. Mag. 58: 641–648.

Bigham, J. M., Schwertmann, U., Carlson, L. and Murad, E. (1990) A poorly crystallized oxyhydroxysulfate of iron formed by bacterial oxidation of Fe(II) in acid mine waters. Geochim. Cosmochim. Acta 54: 2743–2758.

Bigham, J. M., Schwertmann, U., Traina, R. L., Winland, R. L. and Wolf, M. (1996) Schwertmannite and the chemical modelling of iron in acid sulfate waters. Geochim. Cosmochim. Acta 60: 2111–2121.

Binnig, G., Quate, C. F. and Gerber, C. (1986) Atomic force microscope. Phys. Rev. Lett. 26(9), 930–933.

Bish, D. L. and Post, J. E. (eds.) (1989) Modern powder diffraction. Reviews in Mineralogy 20, Min. Soc. Amer. 369 pp.

Bottero J. Y., Manceau A., Villieras, F. and Tchouba, D. (1994) Structure and mechanisms of formation of FeOOH(Cl) polymers. Langmuir 10: 316–319.

Böhm, J. (1925) Über Aluminium- und Eisenoxide I. Z. Anorg. Allg. Chem. 149: 203.

Braun, H. and Gallagher, K. J. (1972) β-Fe_2O_3, a new structural form of iron(III) hydroxide. Nature 240: 13–14.

Brauer, G. (1954) Handbuch der präparativen anorganischen Chemie. F. Enke, Stuttgart.

Breulmann, M., Cölfen, H., Hentze, H.-P.; Antonietti, M., Walsh, D. and Mann, S. (1998) Elastic Magnets: Template controlled mineralization of iron oxide colloids in a sponge-like gel matrix. Advanced Materials 10: 237–241.

Brindley, G. W. and Brown, G. (eds.) (1980) Crystal structures of clay minerals and their X-ray diffraction. Min. Soc. London, 495 pp.

Bruun Hansen H. C., Borggard, O. K. and Sorensen, J. (1994) Evaluation of the free energy of formation of Fe(II)-Fe(III) hydroxide-sulfate (green rust) and its reduction of nitrite. Geochim. Cosmochim. Acta 58: 2599–2608.

Burns, R. G. (1993) Mineralogical applications of crystal field theory. Cambridge University Press

Bürgisser, C., Scheidegger, A. M., Borkovec, M., and Sticher, H. 1994. Chromatographic charge density determination of materials with low surface area. Langmuir 10:855–860.

Büttner, G. (1961) Über Eisen(III)oxid und eine neue Eisenoxidphase ε-Fe_2O_3. Bergakademie 12: 780–781.

Buxbaum, G. (Ed.) (1993) Industrial inorganic pigments. VCH Weinheim, 281 pp.

Cambier. P. (1986) Infrared study of goethite of varying crystallinity and particle size. I Interpretation of OH and lattice vibration frequencies. Clay Min. 21: 191–200.

Carrazana-Garcia, J. A., Loez Quinlela, M. A., and Rivas Rey, J. (1997) Characterization of ferrite particles synthesized in the presence of celluse fibres. Coll. Surf. 121: 61–66.

Carlson, L. and Schwertmann, U. (1990) The effect of CO_2 and oxidation rate on the formation of goethite versus lepidocrocite from an Fe(II)system at pH 6 and 7. Clay Min. 25: 65–7.

Celis, R., Cornejo, J. and Hermosin, M. C. (1998) Textural properties of synthetic clay-ferrihydrite associations. Clay Min. 33: 395–408.

Cheneac, C., Tronc, E. and Jolivet, J. P. (1995) Nanostructured Materials 6: 715.

Chenavas, J., Joubert, J. C., Capponi, J. J. and Marezio, M. (1973) Syntheses des nouvelles phases dense d'oxyhydroxydes M^{3+}OOH des metaux de la première série de transition en milieu hydrothermal et haute pression. Solid State Chem. 6: 1–15.

Chukhrov, F. V., Zvyagin, B. B., Gorshkov, A. I., Ermilova, L.P. and Balashova, V. V. (1973) Ferrihydrite. Izvest. Akad. Nauk. SSSR, Ser. Geol. 4: 23–33.

Chukhrov, F. V., Zvyagin, B. B, Gorshkov, A. I., Ermilova, L. P., Korovushkin, V. V., Rudnitskaya, Y.E. and Yakubovskaya, N.Y. (1977) Feroxyhyte, a new modification of FeOOH. Int. Geol. Rev. 19: 873–890.

Cornell, R. M. (1983) The film forming abilities of iron oxides and oxyhydroxides. Clay Min. 18: 209–213.

Cornell, R. M. (1985) Effect of simple sugars on the alkaline transformation of ferrihydrite into goethite and hematite. Clays Clay Min. 33: 219–227.

Cornell, R. M. (1992) Preparation and properties of Si substituted akaganeite. Z. Planzenernähr. Bodenk. 155: 449–453.

Cornell, R. M. and Giovanoli, R. (1985) Effect of solution conditions on the proportion and morphology of goethite formed from ferrihydrite. Clays Clay Min. 33: 424–432.

Cornell, R. M. and Giovanoli, R. (1986) Factors that govern the formation of multi-domainic goethites. Clays Clay Min. 34: 557–564.

Cornell, R. M. and Giovanoli, R. (1987) Effect of manganese on the transformation of ferrihydrite into goethite and jacobsite in alkaline media. Clays Clay Min. 35: 11–20.

Cornell R. M. and Giovanoli R. (1987 a) The influence of silicate species on the morphology of goethite grown from ferrihydrite. Chem. Comm. 413–414.

Cornell, R. M. and Giovanoli, R. (1988) The influence of copper on the transformation of ferrihydrite ($Fe_2O_3 \cdot 9H_2O$) into crystalline products in alkaline media. Polyhedron 7: 385–391.

Cornell, R. M and Giovanoli R. (1993) Acid dissolution of hematites of different morphologies. Clay Min.28: 223–232.

Cornell, R. M.; Giovanoli, R. and Schindler, P. W. (1987) Effect of silicate species on the transformation of ferrihydrite into goethite and hematite in alkaline media. Clays Clay Min. 35: 21–28.

Cornell, R. M., Giovanoli, R. and Schneider, W. (1989) Review of the hydrolysis of iron (III) and the crystallization of amorphous iron (III) hydroxide hydrate. J. Chem. Technol. Biotechnol. 46: 115–134.

Cornell, R. M., Giovanoli, R. and Schneider, W. (1990) Effect of cysteine and manganese on the crystallization of noncrystalline iron(III)hydroxide at pH 8. Clays Clay Min. 38: 21–28.

Cornell, R. M., Giovanoli, R. and Schneider, W. (1991) Preparation and characterization of colloidal α-FeOOH with a narrow size distribution. J. Chem. Soc. Faraday Trans. 87: 869–873.

Cornell, R. M., Posner, A. M. and Quirk. J. P. (1974) Crystal morphology and the dissolution of goethite. J. inorg. nucl. Chem. 36: 1937–1946.

Cornell R. M. and Schindler, P. W. (1987) Photochemical dissolution of goethite in acid/oxalate solution. Clays Clay Min. 35: 347–352.

Cornell, R. M., Schneider, W. and Giovanoli, R. (1989) Phase transformation in the ferrihydrite/cysteine system. Polyhedron 8: 2829–2836.

Cornell, R. M. and Schwertmann, U. (1996) The Iron Oxides Wiley-VCH 573 pp.

Crosa, M., Boero, V. and Franchini-Angela, M. (1999) Determination of mean crystallite dimensions from X-ray diffraction peak profiles: A comparative analysis of synthetic hematites. Clays Clay Min. 47: 742–747.

David, I. and Welch, J. E. (1956) The oxidation of magnetite and related spinels. Constitution of γ-Fe_2O_3. Trans. Faraday Soc. 52: 1642.

DeGrave, E., Bowen, L. H. and Weed, S. B. (1982) Mössbauer study of aluminum-substituted hematites. J. Magnetism Magnetic Mat. 27: 98–108.

Dezsi, I. and Coey, J. M. D. (1973) Magnetic and thermal properties of α-Fe_2O_3. Phys. Stat. Sol. 15: 681–685.

Drits, V. A., Sakarov B. A. and Manceau, A. (1993) Structure of feroxyhyte as determined by simulation of X-ray diffraction curves. Clay Min. 28: 209–229.

Ebinger, M. H. and Schulze, D. G. (1989) Mn-substituted goethite and Fe-substituted groutite synthesized at acid pH. Clays Clay Min. 37: 151–156.

Ebinger, M. H. and Schulze, D. G. (1990) The influence of pH on the synthesis of mixed Fe-Mn oxide minerals. Clay Min. 25: 507–518.

Eggleston, C. (1994) High resolution scanning microscope: Tip-surface interaction, artifacts and applications in mineralogy and geochemistry. In K. L. Nagy and A. E. Blum (Eds.) Scanning probe microscopy of clay minerals, CMS workshop lectures, Vol. 7, 1–90. The Clay Min. Soc. Boulder, CO.

Eggleton, R. A. and Fitzpatrick, R. W. (1988) New data and a revised structural model for ferrihydrite. Clays and Clay Min. 36: 111–124.

Ennas, G., Musinu, A., Piccaluga, G., Zedda, D., Galleschi, D., Sangregorio, G., Stanger, J. L., Concao, G. and Spano, G. (1998) Characterization of iron oxide nanoparticles in an Fe_2O_3-SiO_2 composite prepared by a sol-gel method. Chem. Mat. 10: 495–502.

Farmer, V. C. (1974) The infrared spectra of minerals. Min. Soc. London, 539 pp.

Fassbinder, J. W. E., Stanjek, H. and Vali, H. (1990) Occurrence of magnetic bacteria in soil. Nature 343: 161–163.

Feitknecht, W. (1959) Über die Oxidation von festen Hydroxyverbindungen des Eisens in wässrigen Lösungen. Z. Elektrochem. 63: 34–43.

Feitknecht, W. and Keller, G. (1950) Über die dunkelgrünen Hydroxyverbindungen des Eisens. Z. Anorg. Chem. 262: 61–68.

Feitknecht, W. and Michaelis, W. (1962) Über die Hydrolyse von Eisen(III)-perchlorat-Lösungen. Helv. Chim. Acta 45: 212–224.

Fernandez R. N. and Schulze D. G. (1987) Calculation of soil color from reflectance spectra. Soil Sci. Soc. Amer. J. 51: 1277–1282.

Finch, G. I. and Sinha, K. P. (1957) An electron diffraction study of the transformation of γ-Fe_2O_3 to α-Fe_2O_3. Proc. Roy. Soc. A241: 1–8.

Fey, M. V. and Dixon, J. B. (1981) Synthesis and properties of poorly crystalline hydrated aluminous goethites. Clays Clay Min. 29: 91–100.

Fitzpatrick, R. W. (1988) Iron compounds as indicators of pedogenic processes: Examples from the southern hemisphere. In Stucki J. W. et al. (Eds.) Iron in soils and clay minerals. Reidel Publ. Co. Dordrecht Holland, NATO ASI Ser. 217: 351–396.

Fischer, W. R. and Schwertmann, U. (1975) The formation of hematite from amorphous iron(III)hydroxide. Clays Clay Min. 23: 33–37.

Ford, R. G., Kemner, K. M. and Bertsch, P. M. (1999) Influence of sorbate-sorbent interaction on the crystallization kinetics of nickel- and lead-ferrihydrite coprecipitates. Geochim. Cosmochim. Acta 63: 39–48.

Forsythe, J. H., Maurice, P. A. and Hersman, L. E. (1998) Attachment of a Pseudomonas sp. to Fe(III)-(hydr)oxide surfaces. Geomicrobiology 15: 293–308.

Gabriel, U., Gaudet, J. P., Spadini, L., and Charlet, L. (1998) Reactive transport of uranyl in a goethite column: an experimental and modelling study. Chemical Geology 151: 107–128.

Galbraith, S. T:, Baird, T. and Fryer, J. R. (1979) Structural changes in β-FeOOH by radiation damage. Acta cryst. A35, 197–200.

Galvez, N., Barron, V. and Torrent, J. (1999) Effect of phosphate on the crystallization of hematite, goethite, and lepidocrocite from ferrihydrite. Clays Clay Min. 47: 304–311.

Galvez, N., Barron, V. and Torrent, J. (1999 a) Preparation and properties of hematite with structural phosphorous. Clays Clay Min. 47: 375–385.

Garg, A. and Matijevic, E. (1988) Preparation and properties of coated uniform colloidal particles. II. Chromium hydrous oxide on hematite. Langmuir 4: 38–44.

Garg, A. and Matijevic, E. (1988 a). Preparation and properties of coated uniform colloidal particles III.Zirconium hydrous oxide on hematite. J. Coll. Interface Sci. 126: 243–250.

Garrels, R. M. and Christ, C. L. (1965). Solutions, minerals and equilibria. Harper and Row, Publ., New York, 450 pp.

Gasser, U. G., Nüesch, R., Singer, M. A. and Jeanroy, E. (1999) Distribution of manganese in synthetic goethite. Clay Min. 34: 291–299.

Geelhoed, J. S., Findenegg, G. R., and van Riemsdijk, W. H. (1997) Availability to plants of phosphate adsorbed on goethite: experiment and simulation. Eur. J. Soil Sci. 48: 473–481.

Gehring, A. U. and Hofmeister, A. M. (1994) The transformation of lepidocrocite during heating: a magnetic and spectroscopic study. Clays Clay Min. 42: 409–415.

Geelhoed, J. S., Findenegg, G. R. and van Riemsdijk, W. H. (1997) Availability to plants of phosphate adsorbed to goethite: experiment and simulation. Eur. J. Soil Sci. 48: 473–481.

Genin, J. R., Bourrie, G., Trollard, F., Abdelmoula, M., Jaffrezic, A., Refait, P., Maitre, V., Humbert, B. and Herbillon, A. (1998) Thermodynamic equilibria in aqueous suspensions of synthetic and natural Fe(II)-Fe(III) green rusts : Occurrences of the mineral in hydromorphic soils. Env. Sci. Techn. 32: 1058–1068.

Gerth, J. (1990) Unit-cell dimensions of pure and trace metal-associated goethites. Geochim. Cosmochim. Acta 54: 363–371.

Giovanoli, R. and Brütsch, R. Dehydration of γ-FeOOH: Direct observation of the mechanism. Chimia 28: 181–184.

Giovanoli R. and Cornell, R. M. (1992) Crystallization of metal-substituted ferrihydrites. Z. Pflanzenern. Bodenk. 155: 455–460.

Glasauer, S. Friedl, J. and Schwertmann, U. (1999) Properties of goethites prepared under acid and basic conditions in the presence of silicate. J. Coll. Interface Sci. 216: 106–115.

Glemser, O. and Gwinner, E. (1939) Über eine neue ferromagnetische Modifikation des Eisen(III)-Oxydes. Z. Anorg. Chem. 240: 161–166.

Goodman, B. A. and Lewis, D. G. (1981): Mössbauer spectra of aluminous goethites (α-FeOOH). J. Soil Sci. 32: 351–363.

Gregg, S. J. and Sing, K. S. W. (1991) Adsorption, surface area and porosity. 2nd ed., Academic Press, London, 371 pp.

Griffith, P. R. and de Haseth, J. A. (1986) Fourier transform infrared spectroscopy. Wiley, 656 pp.

Golden, D. C. D., Ming, W., Bowen, L. H., Morris, R. V. and Lauer, Jr. H. V. (1994) Acidified oxalate and dithionite solubility and color of synthetic, partially oxidized Al-magnetites and their thermal oxidation products. Clays Clay Min. 42: 53–62.

Hahn, F. L. and Hertrich, A. (1923) Leicht filtrierbares Eisenhydroxyd durch Fällung mit Thiosulfat und Jodat. Ber. Dt. Chem. Ges. 26: 1729–1732.

Hamada, S. and Matijevic, E. Ferric hydrous oxide sols. IV. Preparation of uniform cubic hematite particles by hydrolysis of ferric chloride in alcohol-water solutions. J. Coll. Int. Sci. 84: 274–277.

Helgeson, H. C. (1969) Thermodynamics of hydrothermal systems at elevated temperatures and pressures. Amer. J. Sci. 167: 729–804.

Hochella, M. F. jr. (1990) In: Mineral Surfaces, Vaughan, D. J. and Patrick, R. D. (eds.); Chapman and Hall, London.

Huguenin R. L. and Jones J. L. (1986) Intelligent information extraction from reflectance spectra: absorption band positions. J. Geophys. Res. 91, 9585–9598.

Hund F. (1981) Inorganic pigments: Basis for coloured, uncoloured and transparent properties. Angew. Chem. Int. Ed. Eng. 20, 723–730.

Hunter R. S. and Harold R. W. (1987) The measurement of appearance. John Wiley and Sons.

Iul Haq, I. and Matijevic, E. (1997) Preparation and properties of uniform, coated inorganic colloidal particles. X. Manganese compounds on haematite. J. Coll. Interface Sci. 192: 104–113.

Jambor J. L. and Dutrizac, J. E. (1998) Occurrence and constitution of natural and synthetic ferrihydrite, a widespread iron oxyhydroxide. Chem. Rev. 98: 2549–2585.

Janney, D. E., Cowley. J. M. and Buseck, P. R. (2000) Transmission electron microscopy of synthetic 2- and 6-line ferrihydrite. Clays Clay Min. 48: 111–119.

Joekes, I., Galembeck, F., Santos, H. S. and Jafelicci Jr., M. (1981) Preparation and characterization of mono-dispersed iron(III)hydroxide-aqueous ethanolic systems. J. Coll. Interface Sci. 84: 278–280.

Joint Committee on Powder Diffraction Standards. Mineral powder diffraction file. Int. Centre for Diffraction Data, 1601 Park Lane, Swarthmore, Pennsylvania 19081, USA, 1168 pp.

Johnston, J. H. and Lewis, D. G. (1983) A detailed study of the transformation of ferrihydrite to. hematite in an aqueous medium at 92 °C. Geochim. Cosmochim. Acta 47: 1823–1831.

Johnston, J. H. and Lewis, D. G. (1986) A study of the initially-formed hydrolysis species and intermediate polymers and their role in determining the product iron oxides formed in the weathering of iron. In: Long, G. J. and Stevens, J. G. (eds.) Industrial application of the Mössbauer effect.. Plenum, New York, 565–583.

Kandori, K., Uchida, S., Kataoka, S. and Ishikawa, T. (1992) Effects of silicate and phosphate ions on the formation of ferric hydroxide particles. J. Mat. Sci. 27: 719–728.

Kandori, K., Kawashima, Y. and Ishikawa, T. (1993) Zeta potential of monodispersed hematite paticles containing chloride ions J. Mat. Sci. Letters 12: 288–299.

Kandori, K., Horii, I, Yasukawa, A. and Ishikawa, T. (1995) Effects of surfactants on the precipitation and properties of colloidal particles from forced hydrolysis of $FeCl_3$–HCl solution. J. Mat. Sci. 30: 2145–2152.

Kandori, K., Yasukawa, A. and Ishikawa, T. (1996) Influence of amines on formation and texture on uniform hematite particles. J. Coll. Interface Sci. 180: 446–452

Kandori, K., Nakamoto, Y. Yasukawa, A. and Ishikawa, T. (1998) Factors in the precipitation medium governing morphology and structure of hematite particles in forced hydrolysis reactions. J. Coll. Interface Sci. 202: 499–506.

Kandori, K., Ohkoshi, N., Yasukawa, A. and Ishikawa, T. (1998 a) Morphology control and texture of hematite particles by dimethylformamide in forced hydrolysis reaction. J. Mat. Res. 13: 1698–1706

Karim, Z. (1984) Characteristics of ferrihydrite formed by oxidation of $FeCl_2$ solutions containing different amounts of silica. Clays Clay Min. 32: 181–184

Kauppinen J. K., Moffatt D. J., Mantsch H. H., and Cameron D. G. (1981) Fourier self-deconvolution: A method for resolving intrinsically overlapped bands. Appl. Spectroscopy 35: 271–276.

Kiyama, M., Akita, T., Shimizu, S., Okuda, Y. and Takada, T. (1972) Conditions favorable for the formation of γ-FeOOH by aerial oxidation in an acid suspension of iron metal powder. Bull. Chem. Soc. Japan 45: 3422–3426.

Klug, H.P . and L. E. Alexander (1974) X-ray diffraction procedures. J. Wiley and Sons Inc., New York, 555 pp.

Koch, C. J. W, Borgaard, O. K, Madsen, M. B. and Mørup, S. (1987) Magnetic properties of synthetic feroxyhyte (δ′-FeOOH). In: Schultz LG (Ed.) Proc. Int. Clay Conf. Denver 1985: 212–220.

Kosowski, B. M. (1993) Nanometer sized iron oxide. Paper presented at: Iron oxides in colorant and chemical application. Intertec Conferences. Washington DC.

Kratohvil S. and Matjevic, E. (1987) Preparation and properties of coated uniform colloidal particles. I. Aluminum (hydrous) oxide on hematite, chromia and titania. Adv. Ceramic Mat. 2: 798–803.

Krishnamurti G. S. R. and Huang, P. M. (1993) Formation of lepidocrocite from iron(II) solutions. Soil Sci. Soc. Amer. J. 57: 861–867.

Kubelka P. and Munk F. (1931) Ein Beitrag zur Optik der Farbanstriche. Z. Tech. Phys. 11, 593- 603.

Langmuir, D. (1969) The Gibbs free energy of substances in the system Fe-O_2-H_2O-CO_2 at 25 °C. U.S. Geol. Surv. Prof. Paper 650-B: B180-B184.

Langmuir, D. (1971) Particle size effect on the reaction goethite = hematite + water Amer. J. Sci. 271: 147–151.

Lewis, D. G. (1997) Factors influencing the stability and properties of green rusts. Adv. Geoecology 30: 345–347.

Lewis, D. G. and Schwertmann, U. (1979) The influence of aluminum on iron oxides. IV. The influence of [Al], [OH], and temperature. Clays Clay Min. 27: 195–200.

Lewis, D. G. and Schwertmann, U. (1980) The effect of [OH] on the goethite produced from ferrihydrite under alkaline conditions. J. Coll. Interface Sci. 78: 543–553.

Lim-Nunez, R. and Gilkes, R. J. (1989) Acid dissolution of synthetic metal-containig goethites and hematites. In: Schultz, L. G., van Olphen, H. and Mumpton, F. A. (eds.), Proc. Int. Clay Conf. Denver, 1985, Clay Min. Soc., Bloomington, IN, 197–204.

Lowenstam, H. A. and Kirschvink, J. L. (1985) Iron biomineralization, a geobiological perspective. In: Kirschvink J.L., Jones, D. S. and MacFadden, B. J. (eds.), Magnetite Mineralization and Magnetoreception in Organisms. Plenum Publ. Co., 3–15.

Mackay, A. L. (1961) Some aspects of the topochemistry of the iron oxides and hydroxides. In: deBoer, J.H. (ed.). Reactivity of Solids. Proc. 4th Intern. Symposium on the Reactivity of Solids, 571–583.

Mackenzie, R. C. (ed.) (1957) The differential thermal investigation of clays. Min Soc. London. 456 pp.

Mann, S., Skarnulis, A. S. and Williams, R. J. P. (1979) Location of biological compartments by high resolution NMR spectroscopy and electron microscopy using magnetite containing vesicles. J. Chem. Soc., Chem. Comm. 1067–1068.

Mann, S., Webb, J. and Williams, R. J. P. (1989) Biomineralization: Chemical and biochemical perspectives. VCH Weinheim, 548 pp.

Matijevic, E. and Sapiesko, R. S. (2000) Metal oxides: Forced hydrolysis in homogeneous solutions. In: T. Sugimoto (ed.), Fine particles: Synthesis, characterization and mechanism of growth. Marcel Dekker, New York, in press.

Matijevic, E. and Scheiner, P. (1978) Ferric hydrous oxide sols. III Preparation of uniform particles by hydrolysis of Fe(III) chloride – nitrate and perchlorate solutions. J. Coll. Interface Sci. 63: 509–524.

Matijevic, E. and Partsch, R. E. (1999) Synthesis of monodispersed colloids by chemical reactions. In: Sugimoto, T. (ed.). Fine Particles: Synthesis, Characterization and Mechanism of Growth. Marcel Dekker, New York , in press

Maurice, P. A. (1998) Scanning probe microscopy of environmental surfaces. In: Huang, P. M. et al. (Eds.) Structure and surface reactions of soil particles. 109–153.

Maurice, P. A., Hochella, M. F. jr., Parks, G. A., Sposito, G. and Schwertmann, U. Evolution of hematite surface microtopography upon dissolution by simple organic acids. Clays Clay Min. 43: 29–38.

Maurice, P. A., Forsythe, J., Hersman, L. and Sposito, G. (1996) Application of atomic-force electron microscopy to studies of microbial interactions with hydrous Fe(III)-oxides. Chem. Geol.132: 33–43.

McMillan, S. G. and Schwertmann, U. (1998) Morphological and genetic relations between siderite, calcite and goethite in a low moor peat of southern Germany. Eur. J. Soil Sci. 49: 283–294.

Meeussen, J. C. L., Scheidegger, A. M., Hiemstra, T., van Riemsdijk W. H., and Borkovec, M. (1996). Predicting multicomponent adsorption and transport of fluoride at variable pH in a goethite – silica sand system. Environ. Sci. Technol. 30: 481–488.

Meeussen, J. C. L., Kleikemper, J., Scheidegger, A. M., Borkovec, M, Paterson, E., van Riemsdijk W. H., and Sparks, D. L. (1999) Multicomponent transport of sulfate in a goethite-silica sand system at variable pH and ionic strength. Environ. Sci. Techn. 33: 3443–3450.

Mohr, E. C. J.; van Baren, F. A.; van Schuylenbourg, J. (1972) Tropical Soils. 3rd ed. Monton-Ichtior, Baru, van Hoeve, The Hague-Paris-London. 481 pp.

Morris, R. V., Agresti, D. G., Lower, H. V., Newcomb, J. A., Shelfer T. D. and Murali, A. V. (1989) Evidence of pigmentary hematite on Mars based on optical, magnetic, and Mössbauer studies of superparamagnetic (nanocrystalline) hematite. J. Geophys. Res. 94: B32760–2778.

Morris, R. V., Lauer, H. V. jr., Schulze D. G. and Burns, R. G. (1991) Preparation and characterization of nanophase hematite powder. Lunar Plenetary Sci. 22: 927–928.

Mørup, S., Madsen, M. B., Franck, J.,Villadsen, J. and Koch, C. J. W. (1983) A new interpretation of Mössbauer spectra of microcrystalline goethite: "super-ferromagnetic" or "super-spin-glass" behavior? J. Magn. magn. Mat. 40: 163–174.

Mullin, J. W. (1993) Crystallization, 3rd edition. Butterworth-Heinemann, pp 527.

Murad, E. (1979) Mössbauer and X-ray data on β-FeOOH (akaganeite). Clay Min. 14: 273–283.

Murad, E. (1996) Magnetic properties of microcrystalline iron(III)oxides and related minerals as reflected in their Mössbauer spectra. Phys. Chem. Min. 23: 248–262

Murad, E., Bowen, L. H.; Long, G. L. and Quin, T. G. (1988) The influence of crystallinity on magnetic ordering in natural ferrihydrites. Clay Miner. 23: 161–173.

Murad, E. and Johnston, J. H. (1987) Iron oxides and oxyhydroxides. In: Long, G. (Ed.) Mössbauer spectroscopy applied to inorganic chemistry. Plenum Publ. Corp., New York, 2: 507–582.

Murad, E. and Schwertmann, U. (1980) The Mössbauer spectrum of ferrihydrite and its relations to those of other iron oxides. Amer. Miner. 65: 1044–1049.

Murad, E. and Schwertmann, U. (1983) The influence of aluminium substitution and crystallinity on the Mössbauer spectra of goethite. Clay Min. 18: 301–312.

Murad, E. and Schwertmann, U. (1984) The influence of crystallinity on the Mössbauer spectrum of lepidocrocite. Mineral. Magazine 48: 507–511.

Murad E. and Schwertmann, U. (1986) Influence of Al substitution and crystal size on the room-temperature Mössbauer spectrum of hematite. Clays Clay Min. 34: 1–6.

Murad, E. and Schwertmann, U. (1993) Temporal stability of a fine-grained magnetite. Clays Clay Min. 41: 111–113.

Murad, E. and Taylor, R. M. (1984) The Mössbauer spectra of hydroxy carbonate green rust. Clay Min. 19: 77–83.

Murad, E. and Taylor, R. M. (1986) The oxidation of the hydroxycarbonate green rusts. In: Long, G. J. and Stevens J. G. (eds.) Industrial applications of the Mössbauer effect. Plenum. Publ. Corp. 585–593.

Muramatsu, A., Ishikawa, S. and Sugimoto, T. (1994) Controlled formation of ultrafine nickel particles on well-defined hematite particles. Coll. Surf. A 82: 29–35.

Murray, J. W. (1979) Iron oxides. In: Burns, R. G. (ed) Marine minerals. Min. Soc. Amer. Short Course Notes, Vol. 6, Washington D.C., 47–98.

Naono, H., Sonoda, J., Oka, K. and Hakuman, M. (1993) Evaluation of microporous texture of undecomposed and decomposed β-FeOOH fine particles by means of adsorption isotherms of nitrogen gas and water vapor. Proc. IVth Int. Conf. on Fundamentals of Adsorption, Kyoto 1992, 467–474.

Nassau, K. (1983) The Physics and Chemistry of Color: The Fifteen Causes of Color. Wiley & Sons, New York, 454 pp.

Nielson, A. E. (1964) Kinetics of Precipitation, Pergamon, Oxford.

Ohmori, M. and Matijevic, E. (1992) Preparation and properties of uniform coated colloidal particles. VII. Silica on hematite. J. Coll. Interface Sci. 150: 594–598.

Ozaki, M., Kratohvil, S. and Matijevic, E. (1984) Formation of monodispersed spindle type haematite particles. J. Coll. Interface Sci. 102: 146–151.

Ozaki, M. and Matijevic E. (1985) Preparation and magnetic properties of monodispersed spindle-type γ-Fe_2O_3 particles. J. Coll. Interface Sci. 107: 199–203.

Paterson, E. and Tait, J. M. (1977) Nitrogen adsorption on synthetic akaganeite and its structural implications. Clay Min. 12: 345–352.

Philipse, A. P., van Bruggen, M. P. B.and Pathmannanoharan, C. (1994) Magnetic silica dispersions: preparation and stability of surface modified silica particles with a magnetic core. Langmuir 10: 92–97.

Post D. F., Levine S. J., Bryant R. B., Mays M. D., Batchily A. K., Escadafal R., and Huete A. R. (1993) Correlations between field and laboratory measurements of soil color. In Soil Color. In: Bigham J. M. and Ciolkosz E. (Eds.), Soil Science Society of America, Spec. Publ. No 31: 35–49.

Refait, Ph. and Genin, J.-M. R. (1997) Mechanism of oxidation of Ni(II)-Fe(II) hydroxides in chloride-containig aqueous media: role of the pyroaurite-type Ni-Fe hydroxychloride. Clay Min. 32: 597–613.

Raven, K. P., Jain, A. and Loeppert, R. H. (1998) Arsenite and arsenate adsorption on ferrihydrite: Kinetics, equilibrium, and adsorption envelopes. Env. Sci. Technol. 32: 344–349.

Riveros, P. A. and Dutrizac, J. E. (1997) The precipitation of hematite from ferric chloride media. Hydrometallurgy 46: 85–104.

Robie, R. A. and Waldbaum, D. R. (1967) Thermodynamic properties of minerals and related substances at 298.15°K (25 °C) and one atmosphere (1.013 bars) pressure and at higher temperatures. Geol. Surv. Bull. (U.S.) 1259.

Rose, J., Manceau, A., Masion, A., and Bottero, J.-Y. (1997) Structure and mechanisms of formation of $FeOOH(NO_3)$ oligomers in the early stage of hydrolysis. Langmuir 13: 3240–3246.

Russell, J. D. (1979) Infrared spectroscopy of ferrihydrite: evidence for the presence of structural hydroxyl groups. Clay Miner. 14: 190–214.

Sapieszko, R. S. and Matijevic, E. (1980) Preparation of well defined colloidal particles by thermal decomposition of metal chelates. I Iron oxides. J. Coll. Interface Sci. 74: 405–422.

Scheidegger, A., Borkovec, M. and Sticher, H. (1993) Coating of silica sand with goethite: Preparation and analytical identification. Geoderma, 58: 43–65.

Scheidegger, A. M., Bürgisser, C., Borkovec, M., Sticher, H., Meeussen, H. and van Riemsdijk W. H. (1994) Convective transport of acids and bases in porous media. Water Resour. Res. 30: 2937–2944.

Scheinost A. C., Chavernas A., Barrón, V., and Torrent J. (1998) Use and limitations of second-derivative diffuse reflectance spectroscopy in the visible to near-infrared range to identify and quantify Fe oxide minerals in soils. Clays Clay Miner. 46, 528–536.

Scheinost A. C., Schulze D. G., and Schwertmann U. (1999) Diffuse reflectance spectra of Al substituted goethite: A ligand field approach. Clays Clay Miner. 47, 156–164.

Scheinost A. C. and Schwertmann U. (1997) VIS-NIR reflectance spectra of goethite (α-FeOOH) as a function of particle size, unit-cell size, and cation substitution. 28th Lunar and Planetary Science Conference, March 17–21, Houston, TX.

Scheinost A. C. and Schwertmann U. (1999) Color identification of iron oxides and hydroxysulfates – Use and limitations. Soil Sci. Soc. Amer. J. 63, 1463–1471.

Schulze, D. G. (1984) The influence of aluminium on iron oxides. VIII. Unit cell dimensions of Al-substituted goethites and estimation of Al from them. Clays Clay Min. 32: 36–44.

Schulze, D. G. and Schwertmann, U. (1984) The influence of aluminium on iron oxides. X. Properties of Al-substituted goethites. Clay Min. 19: 521–529.
Schulze, D. G. and Schwertmann, U. (1987) The influence of aluminium on iron oxides: XIII. Properties of goethites synthesized in 0.3 M KOH at 25 °C. Clay Min. 22: 83–92.
Schwertmann, U. (1959 a) Die fraktionierte Extraktion der freien Eisenoxyde in Böden, ihre mineralogischen Formen und ihre Entstehungsweisen. Z. Pflanzenernähr. Düng. Bodenkd. 84: 194–204.
Schwertmann, U. (1959 b) Über die Synthese definierter Eisenoxyde unter verschiedenen Bedingungen. Z. Anorg. Allg. Chem. 298: 337–348.
Schwertmann, U. (1964) Differenzierung der Eisenoxide des Bodens durch Extraktion mit Ammoniumoxalat-Lösung. Z. Pflanzenern. Düng. Bodenkd. 105: 194–202.
Schwertmann, U. (1965) Zur Goethit- und Hämatitbildung aus amorphem Eisen (III)-hydroxid. Z. Pflanzenern., Düngg., Bodenk. 108: 37–45.
Schwertmann, U. (1973) Use of oxalate for Fe extraction from soils. Canad. J. Soil Sci.53: 244–246.
Schwertmann, U. (1984) The influence of aluminium on iron oxides. IX. Dissolution of Al-goethites in 6 M HCl. Clay Min. 19: 9–19.
Schwertmann, U. (1985) The effect of pedogenic environments on iron oxide minerals. Adv. Soil Sci. 1: 172–200.
Schwertmann, U. (1988) Some properties of soil and synthetic iron oxides. In: J.W. Stucki, B.A. Goodman and U. Schwertmann (eds.). Iron in Soils and Clay Minerals, NATO ASI, 1985, Bad Windsheim, F.R.G., 203–244.
Schwertmann, U. (1988) Occurrence and formation of iron oxides in various pedoenvironments. In: J.W., Stucki, B.A. Goodman and U. Schwertmann (eds.). Iron in Soils and Clay Minerals, NATO ASI, 1985, Bad Windsheim, F.R.G., NATO ASI Ser.C 217: 267–302.
Schwertmann U. (1993) Relations between iron oxides, soil color and soil formation. In: Bigham J. M and Ciolkosz, E. J. (eds), Soil Color; Spec. Publ. No. 31: 51–69. Soil Science Society of America, Madison. Wisconsin.
Schwertmann, U., Cambier, P. and Murad, E. (1985) Properties of goethites of varying crystallinity. Clays Clay Min. 33: 369–378.
Schwertmann, U., Carlson, L., Stanjek, H. and Kämpf, N. (1989) Calibration of Al substitution in goethite and hematite with natural samples. Abstract, Int. Clay Conf. Straßbourg, 347.
Schwertmann, U. and Carlson, L. (1994) Aluminum influence on iron oxide: XVII. Unit cell parameters and aluminum substitution of natural goethites. Soil Sci. Soc. J. 58: 256–261.
Schwertmann, U. and Fechter, H. (1984) The influence of aluminium on iron oxides. XI. Aluminium-substituted maghemite in soils and its formation. Soil Sci. Soc. Amer. J. 48: 1462–1463.
Schwertmann, U. and Fechter, H. (1994) The formation of green rust and its transformation to lepidocrocite. Clay Min. 29: 87–92.

Schwertmann, U., Fechter, H., Taylor, R. M. and Stanjek, H. (1995) A lecture and demonstration for students on iron oxide formation. In: G.J. Churchman, R.W. Fitzpatrick and R.A. Eggleton (Editors) Clays: Controlling the environment. Proc. 10th Int. Clay Conf., Adelaide, Australia, 1993,. CSIRO Publishing, Melbourne, Australia, p. 11–14.

Schwertmann, U. and Fischer, W. R. (1966) Zur Bildung von α-FeOOH und α-Fe_2O_3 aus amorphem Eisen(III)-hydroxid. III. Z. Anorg. Allg. Chem. 346: 137–142.

Schwertmann, U., Fischer, W. R. and Papendorf, H. (1968) The influence of organic compounds on the formation of iron oxides. Trans. 9th Int Congr. Soil Sci. Adelaide, Australia, 1: 645–655.

Schwertmann, U. and Fischer, W. R. (1973) Natural "amorphous" ferric hydroxide. Geoderma 10: 237–247.

Schwertmann, U., Fitzpatrick, R. W., Taylor, R. M. and Lewis, D. G. (1979) The influence of aluminum on iron oxides. Part II. Preparation and properties of Al-substituted hematites. Clays Clay Min. 27: 105–112.

Schwertmann, U. and Friedl, J. (1998) Thin iron oxide films on pebbles in ferriferrous streams. N. Jahrb. Min. Monatshefte 2: 63–67.

Schwertmann, U., Friedl, J. and Pfab, G. (1996) A new iron(III)hydroxynitrate. J. Solid State Chem. 126: 336.

Schwertmann, U., Friedl, J. and Stanjek H. (1999) From Fe(III) ions to ferrihydrite and then to hematite. J. Coll. Interface Sci. 209: 215–223.

Schwertmann, U., Friedl, J., Stanjek, H. and Schulze, D. G. (2000) The effect of Al on Fe oxides. XIX. Formation of Al.substituted hematite from ferrihydrite at 25 °C and pH 4 to 7. Clays Clay Min. 48: in press.

Schwertmann, U., Friedl, J. and Lewis D. G. (2000 a) Synthetic ferrihydrites of different crystallinity and their relation to FeOOH forms. in prep.

Schwertmann, U., Gasser, U. and Sticher, H. (1989) Chromium-for-iron substitution in synthetic goethites. Geochim. Cosmochim. Acta 53: 1293–1297.

Schwertmann, U. and Murad, E. (1983) Effect of pH on the formation of goethite and hematite from ferrihydrite. Clays Clay Min. 31: 277–284.

Schwertmann, U. and Murad, E. (1990) The influence of aluminum on iron oxides: XIV. Al-substituted magnetite synthesized at ambient temperatures. Clays Clay Min. 38: 196–202.

Schwertmann, U. and Pfab, G. (1994) Structural vanadium in synthetic goethite Geochim. Cosmochim. Acta 58:4349–4352.

Schwertmann, U. and Stanjek, H. (1998) Stirring effects on properties of Al goethite formed from ferrihydrite. Clays Clay Min. 46: 317–321.

Schwertmann, U. and Taylor, R. M. (1989) Iron oxides. In: Dixon, J. B. and Weed, S.B. (Eds.) Minerals in soil environments. 2nd. ed. Soil Sci. Soc Amer. Book Ser. No 1 Madison Wisc. 379–438.

Schwertmann, U. and Thalmann, H. (1976) The influence of Fe(II), Si and pH on the formation of lepidocrocite and ferrihydrite during oxidation of aqueous $FeCl_2$ solutions. Clay Min. 11: 189–200.

Schwertmann, U. and Wolska, E. (1990) The influence of aluminum on iron oxides. XV. Al-for-Fe substitution in synthetic lepidocrocite. Clays Clay Min. 38: 209–212.

Sidhu, P. S., Gilkes, R. J. and Posner, A. M. (1977) Mechanism of the low temperature oxidation of synthetic magnetites. J. Inorg. Nucl. Chem. 39: 1953–1958.

Sidhu, P. S., Gilkes, R. J. and Posner, A. M. (1978) The synthesis and some properties of Co, Ni, Zn, Cu, Mn and Cd substituted magnetites. J. Inorg. Nucl. Chem. 40: 429–435.

Sidhu, P. S., Gilkes, R. J., Cornell, R. M., Posner, A. M. and Quirk, J.P . (1981) Dissolution of iron oxides and oxyhydroxides in hydrochloric and perchloric acids. Clays Clay Min. 29: 269–276.

Sinke, A. J. C., Dury, O., and Zobrist, J. (1998) Effects of a fluctuating water table: column study on redox dynamics and fate of some organic pollutants. J. Contam. Hydrol. 33: 231–246.

Skinner, H. C. W, and Fitzpatrick R. W. (eds.) (1992) Biomineralization processes of iron and manganese. Catena Suppl. 21, Catena Verlag, Cremlingen-Drestedt, Germany; 432 pp.

Smykatz-Kloss, W. (1974) Differential thermal analysis. Springer Verlag, Berlin, 185 pp.

Sorum, S. H. (1928) Preparation of chloride free, colloidal Fe_2O_3 from $FeCl_3$. J. Amer. Chem. Soc. 1263–1267.

Stanjek, H. (1991) Aluminium- und Hydroxylsubstitution in synthetischen und natürlichen Hämatiten. Dissertation, Techn. Univ. München, 194 S.

Stanjek, H. and Schwertmann, U. (1992) The influence of aluminun on iron oxides Part XVI: Hydroxyl and aluminum substitution in synthetic hematites. Clays Clay Min. 40: 347–354.

Stanjek, H., Friedl, J. and Schwertmann, U. (1999) OH substitution in maghemites as an indicator for burning in soils (in prep.).

Stiers, W. and Schwertmann, U. (1985) Evidence for manganese substitution in synthetic goethite. Geochim. Cosmochim. Acta 49: 1909–1911.

Sugimoto, T., Khan, M. M., Muramatsu, A. and Itoh, H. (1993) Formation mechanism of monodisperse peanut-type α-Fe_2O_3 particles from condensed ferric hydroxide gel. Coll. Surf. A 79: 233–247.

Sugimoto, T. and Matijevic, E. (1980) Formation of uniform spherical particles by crystallization of ferrous hydroxide gels. J. Coll. Interface Sci. 74: 227–243.

Sugimoto, T., Waki, S., Itoh, H. and Muramatsu, A. (1996) Preparation of monodisperse plate-let type particles from a highly condensed β-FeOOH suspension. Coll. Surf. 109: 155–165.

Sugimoto, T., Yinsheng, W., Itoh, H. and Muramatsu, A. (1998) Systematic control of size, shape and internal structure of monodispersed α-Fe_2O_3 Coll. Surf. A 134: 265–279.

Sugimoto, T. and Sakata, K. (1992) Preparation of monodispersed pseudocubic α-Fe_2O_3 particles from condensed ferric hydroxide gel. J. Coll. Interface Sci.152: 587–590.

Sunshine J. M., Pieters C. M., and Pratt S. F. (1990) Deconvolution of mineral absorption bands: an improved approach. J. Geophys. Res. 95: 6955–6966.

Sylva, R. M. (1972) The hydrolysis of iron(III). Rev. Pure Appl.Chem. 22: 115–131.

Taylor, R. M. (1980) Formation and properties of Fe(II),(III) hydroxy carbonate and its possible significance in soil formation. Clay Min. 15: 369–382.

Taylor, R. M. (1987) Non-silicate oxides and hydroxides. In: Newman A.C.D. (Ed.) Chemistry of clays and clay minerals. Longman group UK Ltd. Harlow, 129–201.

Taylor, R. M. (1988) Proposed mechanism for the formation of soluble Si-Al- and Fe(III)-Al complexes in soils. Geoderma 42: 65–77.

Taylor R. M, Maher, B. A. and Self, P. G. (1987) Magnetite in soils: I. The synthesis of single-domain and superparamagnetic magnetite. Clay Min. 22: 411–422.

Taylor, R. M. and Schwertmann, U. (1974) Maghemite in soils and its origin. II. Maghemite syntheses at ambient temperature and pH 7. Clay Min. 10: 299–310.

Taylor, R. M., Schwertmann, U. and Fechter, H. (1985) A rapid method for the formation of Fe(II)Fe(III) hydroxy carbonate. Clay Min. 20: 147–151.

Taylor, R. M. and McKenzie R. M. (1980) The influence of aluminium on iron oxides. VI. The formation of Fe(II)-Al(III) hydroxy-chlorides, -sulfates and -carbonates as new members of the pyroaurite group and their significance in soils. Clays Clay Min. 28: 179–187.

Tejedor-Tejedor, M. I. and Anderson, M. A. (1986) In situ attenuated total reflectance Fourier transform infrared studies of the goethite-aqueous solution interface. Langmuir 2: 203–210.

Thiel, R. (1963) Zum System α-FeOOH-α-AlOOH. Z. Anorg. Allg. Chem. 326: 70–78.

Torrent J. and Barrón, V. (1993) Laboratory measurements of soil color: theory and practice. In Bigham, J. M. and. Ciolkosz, E. J. (Eds.) Soil Color, Soil Science Society of America,. Special Publ. No. 31: 21–33.

Torrent J. and Schwertmann U. (1987) Influence of hematite on the color of red beds. J. Sediment. Petrol. 57: 682–686.

Towe, K. M. and Bradley, W. F. (1967) Mineralogical constitution of colloidal "hydrous ferric oxides". J. Coll. Interface Sci. 24: 384–392.

Trolard, F., Genin, J.-M. R., Abdelmoula, M., Bourrie, G., Humbert, B. and Herbillon, A. (1997) Identification of a green rust mineral in a reductive soil by Mössbauer and Raman spectroscopies. Geochim. Cosmochim. Acta 61: 1107–1111.

Tros, M. E., Schraa, G., Zehnder, A. J. B., and Bosma, T. N. P. (1998) Anomalies in the transformation of 3-chlorobenzoate in percolation columns with Pseudomonas SP. strain B13. Wat. Sci. Tech. 37: 89–96.

Uchida, S., Sato, T. and Okuwaki, A. (1993) Synthesis of monodispersed micaceous iron oxide by the oxidation of iron with oxygen under hydrothermal conditions. J. Chem.Techn. Biotechn. 57: 221–227.

Van Schuylenborgh, J. (1973) Sesquioxyde formation and transformation. In: Schlichting, E. and Schwertmann, U. (Eds.) Pseudogley and Gley. Transact. Comm. V and VI, Int. Soc Soil Sci. VCH, Weinheim, 91–102.

Vegard, L. (1921) Die Konstitution der Mischkristalle und die Raumerfüllung der Atome. Z. Physik 5: 17.

Vempati, R. K., Morris, R. H., Lauer jr., H. V. and Helmke, P. A. (1995) Physico-chemical properties of Mn-goethites and hematites J. Geophys. Res. 100: 3285–3295.

Viswanathiah, M. N., Tareen, J. A. K. and Krishnamurthy, K. V. (1980) Low temperature hydrothermal synthesis of magnetite. J. Cryst. Growth 49: 189–192.

Walton, A. G. (1967) The Formation and Properties of Precipitates. Interscience 23, New York.

Waychunas, G. A., Xu, N., Fuller, C. C., Davis, J. A. and Bigham, J. M. (1994) XAS study of AsO_4^{3-} and SeO_4^{2-} substituted schwertmannites. Physica B.

Weidler, P. G. (1997) BET sample pretreatment of synthetic ferrihydrite and its influence on the determination of surface area and porosity. J. Porous Mat. 4: 165–169.

Weidler, P. G., Schwinn, T., and Gaub, H. E. (1996) Vicinal faces on synthetic goethite observed by atomic force microscopy. Clays Clay Min. 44: 437–443.

Weidler, P. G., Degovics, G. and Laggner, P. (1998) Surface roughness created by acidic dissolution of synthetic goethite monitored with SAXS and N_2-adsorption isotherms. J. Coll. Interface Sci. 197: 1–8.

Welo, L. and Baudisch, O. (1925) The two-stage transformation of magnetite into hematite. Phil. Mag. 50: 399–408.

Wilson, M. J. (ed.) (1987) A handbook of determinative methods in clay mineralogy. Blackie, Glasgow and London, 308 pp.

Wolska, E. (1981) The structure of hydrohematite. Z. Krist. 154: 69–75.

Wolska, E. (1988) Relation between the existence of hydroxyl ions in the ionic sublattice of hematite and its infrared and X-ray characteristics. Solid State Ionics 28–30: 1349–1351.

Wolska, E. and Schwertmann, U. (1989) Nonstoichiometric structures during dehydroxylation of goethite. Z. Kristallogr. 189: 223–237.

Wolska, E., Subert, J, Haba, Z., Tlaskal, J. and Schwertmann, U. (1994) X.ray powder diffraction and Mössbauer spectroscopic studies on the solubility limits in α-FeOOH/α-AlOOH solid solutions. J. Mat. Sci. 29: 3269–3273

Wyszecki, G. and Styles, W. S. (1982) Color Science: Concepts and methods, quantitative data and formulae. 2nd ed. Wiley and Sons, New York.

Zedda, D., Gallesch, D., Sangregorco, G., Stanger, J. L., Concao, G. and Spano, G. (1998) Characterization of iron oxide nanoparticles in an Fe_2O_3-SiO_2 composite prepared by a sol-gel method. Chem. Materials 19: 495–502.

Zhao, J., Huggins, F. E., Feng, Z., Lu, F., Shah, N. and Huffman, G. P. (1993) Structure of nano-phase iron oxide catalyst. J. Catalysis 143: 499–509.

Acknowledgement

The following publishers kindly granted permission to reproduce figures:
Academic Press: Figures 2.1, 10.6 (lower)
Catena Verlag GmbH: Figure 13.2
CSIRO Publishing: Figures 16.1, 16.2, Chapter 16
Elsevier Science: Figures 5.16, 10.6 (upper), 14.1, 14.2, 14.3, 14.4, 15.1
Georg Thieme Verlag: Figure 3.6
Leidorf Verlag: Figure 3.10b
Munsell Division of GretagMacbeth: Figure 3.4, Plate of Munsell Colour tree
Plenum Press: Figure 13.1
Soil Science Society of America: Figures 3.3, 5.5
Springer Verlag: Figure 6.3
The Clay Minerals Society: Figures 3.1, 3.8, 3.10 (upper), 5.14, 8.3, 10.5a, b
The International Association for the Study of Clays: Figure 10.5 (in part)
The Mineralogical Society: Figures 5.3, 10.5e
The Royal Society of Chemistry: Figure 5.8
Wiley-VCH Verlag GmbH: Figures 1.1, 5.1, 5.4, 8.2

Index